NICHOLAS COPERNICUS
AND THE FOUNDING
OF MODERN ASTRONOMY

NICHOLAS COPERNICUS
AND THE FOUNDING
OF MODERN ASTRONOMY

Todd Goble

620 South Elm Street, Suite 223
Greensboro, North Carolina   27406
http://www.morganreynolds.com

RENAISSANCE SCIENTISTS

Nicholas Copernicus

Tycho Brahe

Johannes Kepler

Galileo Galilei

Isaac Newton

NICHOLAS COPERNICUS
AND THE FOUNDING OF MODERN ASTRONOMY

Library of Congress Cataloging-in-Publication Data

Goble, Todd, 1962-
  Nicholas Copernicus and the founding of modern astronomy / Todd
Goble.— 1st ed.
     p. cm. — (Renaissance scientists)
Summary: Presents the life and work of the famous sixteenth-century
Polish   astronomer.
  ISBN 1-883846-99-4 (lib. bdg.)
  1. Copernicus, Nicolaus, 1473-1543—Juvenile literature. 2.
Astronomers—Poland—Biography—Juvenile literature. 3.   Copernicus,
Nicolaus, 1473-1543. [1. Astronomers. 2. Scientists.] I. Title. II.
Series.
  QB36.C8G63 2003
  520'.92—dc21

                        2003004659

Printed in the United States of America
First Edition

CONTENTS

Nicholas Copernicus. *(Sixteenth-century woodcut, by Tobias Stimmer.)*

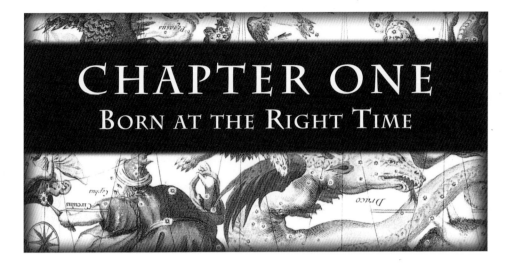

CHAPTER ONE
BORN AT THE RIGHT TIME

One of the most important books in the history of science was almost never published. Nicholas Copernicus worked on the book *On the Revolutions of the Celestial Spheres* for over thirty years without attempting to publish it. The book explained in mathematical detail his theory that the planets, including Earth, orbited the Sun, and that Earth also spun on its axis every twenty-four hours. This theory contradicted the idea that Earth was at the center of the universe, which had been the fundamental idea in astronomy for over two thousand years.

Copernicus knew that to openly question the validity of the Earth-centered universe was risky, and he had never been a risk taker. Although he was one of the most educated men in Europe, Copernicus spent his life working as an official in the Catholic Church. He was highly respected and always carried out his duties in a competent manner, but had not risen to a very high position. Content to watch others assume more power, he preferred spending his time in his tower room, working out the mathematical underpinnings of his planetary model.

Even without publishing the entire book, word of his theory slowly spread among European intellectuals. Copernicus wrote a small pamphlet about his heliocentric ideas that was hand-copied and passed around between the years of 1509 to 1514. The pamphlet was read at some of the universities in Germany, and its readers were interested in seeing the larger manuscript. Many prominent intellectuals, both in and out of the Catholic Church, had written to encourage him to finish and publish the book. Copernicus, however, demurred for many years, never refusing outright to publish, but never committing to doing so either.

No one knows for certain why he was hesitant to let the world read his life's work. It might have been fear of offending the powerful Catholic Church, though that seems unlikely, as many of those encouraging him were church officials. A hundred years would pass before the church would condemn the idea of a moving Earth as contrary to scripture. Most likely, he was aware of problems in his theory that he wanted to fix. He did not

A seventeenth-century illustration of the Copernican system showing the Sun in the center of the universe. *(From Cellarius,* Atlas Coeletis, *1660. Courtesy of the British Library.)*

want to publish until it was as accurate as possible.

Copernicus could have continued working on the manuscript without publishing until his death. It would take a visit in 1539 from a young German scholar who went by the name of Rheticus to convince him to finally release it for publication.

Ironically, Copernicus did not live to see the world's reaction to his life's work. He suffered a stroke weeks

NATI 1473 19 FEBRVARY

Nicholas's father, featured here in this religious painting, was a successful copper merchant in Torun, Poland.

before the book was released and was bedridden until his death. Legend has it he died on the same day he saw the first copy. As with so much about Copernicus's life, we do not know if this legend is true. What we do know is that the influence of *On the Revolutions of the Celestial Spheres* slowly spread across Europe over the next two centuries and eventually led to a complete reevaluation of Earth's place in the universe.

Nicholas Copernicus was born on February 19, 1473, in the town of Torun, Poland, located on the Wisla, a northern flowing river originating south of Cracow. As one of the area's major waterways, the Wisla carried barges and other boats past Torun and Chelmno until it emptied into the Gulf of Gdansk on the Baltic Sea. The Copernicus family moved to Torun from Cracow in the mid-fifteenth century. The astronomer's father, who was also named Nicholas, was a successful copper trader. For the elder Copernicus, settling down in a thriving river-port city like Torun was good for business. Nicholas senior also did well in politics and eventually became a magistrate and an active member of the League of Prussian Towns and Districts, an organization dedicated to Poland's defense.

Copernicus's mother, Barbara Watzenrode, was from a wealthy family. Not much else is known about her, except that she married Nicholas Copernicus around 1463. The Copernicuss moved into a house on St. Anne's Street in one of the wealthier parts of old Torun. Barbara gave birth to four children, two daughters and two sons. Nicholas Copernicus was the youngest child.

The Copernicuss owned a vineyard near the Wisla River on a hill adjacent to a local convent, as well as a summer residence. Winter or summer, city or country-side, young Nicholas witnessed the bustling activity of the cargo vessels hauling ore and other metals and products up and down the river. The state of the economy and trade were the central topics of discussion among his father and his companions. As a child of one of the town's wealthiest families, he was also exposed to music, art, literature, and science from a young age.

Nicholas was fortunate to be born at the height of the Renaissance, a period of rapid intellectual and artistic

PARACELSUS

One of the founders of modern medicine, Theophrastus Phillippus Aureolus Bombastus von Hohenheim, who changed his name to Paracelsus, was born twenty years after Copernicus in what is now Switzerland. The son of a physician and avid alchemist (alchemy is the medieval science and philosophy that aimed to transform elements, such as base metal, into gold, as well as to find a universal cure for illness), Paracelsus learned about medicine and chemistry from his father.

At the age of fourteen, Paracelsus left home and traveled from university to university in search of knowledge and renowned teachers, but always left dissatisfied. During his travels, he often turned to the local midwives, sorcerers, and mystical healers to learn their secrets. He received a medical degree in 1510 from the University of Vienna, and a doctorate in 1516 from the University of Ferrara, where Copernicus also studied. It was around this time that he began calling himself Paracelsus, "para" meaning "above," as a dig at the first century Roman physician Celsus, whose writings were held in high esteem by scholars.

After receiving his degrees, he began extensive travels from Ireland to Turkey. His exciting life included adventures as an army surgeon on military campaigns and escaping from Russian Tartars. Always seeking out new medical and scientific knowledge, he was often more successful than other doctors at treating his patients. He introduced the use of chemicals such as sulfur, mercury, and copper sulfide into medicine.

At the age of twenty-seven, Paracelsus returned to Switzerland and became the town physician and a popular lecturer at the University of Basel. His strong, opinionated personality and vast knowledge led him to criticize current medical practices as well as the philosophies of revered ancient physicians such as the Arab Avicenna and the Greek Galen. He went so far as to publicly burn the written works of these two, and came to be known as the "Martin Luther of medicine." After less than one year in Basel, he had made so many enemies that he had to flee for his life in the middle of the night.

He spent the next eight years accepting the hospitality of friends and writing manuscripts. In 1536, the publication of *Die grosse Wurndartzney* (*Great Surgery Book*) led him to new-found fame. At the age of forty-eight, during the height of his success, he died mysteriously at an inn in Salzburg. It was speculated that he had been poisoned.

development throughout Western Europe. After its inception in Italy, the movement spread throughout Europe and thrived mostly in areas that prospered on trade and commerce. Although situated on what was then the edge of the Christian world, Torun was an ideal city in which the Renaissance could flourish because its extensive trade relationships stretched into Scandinavia and Asia. Copernicus lived at the same time as many of the most familiar names of the Renaissance era, including the artists Michelangelo and Leonardo da Vinci, the writers Machiavelli and Rabelais, and the scientist and alchemist Paracelsus. Most importantly, Copernicus was born twenty years after the invention of movable type, which began the modern publishing industry and made it much easier to disseminate new ideas.

The cultural and scientific achievements of the Renaissance were dependent upon flourishing economies. To help protect its economic development, the town of

LVCAS de AILEN.

After the death of Nicholas's father, his uncle Lucas Watzenrode adopted him. Watzenrode was an important official in the Catholic Church. His influence would help Copernicus be elected as a canon.

Torun had joined the Hanseatic League. Officially organized in 1358, the League developed from a group of merchants who had banded together to fight piracy and foreign competition. Eventually, more than seventy cities were part of the League, from London in the west to Torun in the east. The power these merchants gained by organizing was a development that helped to create the middle class. Instead of scratching a subsistence living from the soil as the peasants before them had done, the new middle class lived in cities and profited from trading goods made by craftsmen. Copernicus grew up in the midst of this urban hustle and bustle, and he learned from the commerce and business that surrounded him.

In 1483, when Nicholas was ten, his father died, and life quickly changed for the young boy. Nicholas and his three siblings came under the care of Barbara's brother, a prominent Catholic Church official named Lucas Watzenrode. It is unclear what happened to Barbara Copernicus. No information about her life after this point exists. Of the children, one of Nicholas's sisters became a nun and the other married a wealthy business-man from Cracow. Uncle Lucas decided to educate the sons, Nicholas and Andreas, together.

Lucas Watzenrode had been born in 1447. After at-tending the University of Cracow at the age of sixteen, he studied at the University of Bologna in Italy and received a doctor's degree in canon law in 1475. Just six years after adopting Nicholas and his siblings, Lucas was appointed the bishop of Varmia in 1489, a position he held until his death.

Uncle Lucas was more than a church official, though, he was also an important political figure in the conflict between the Polish king and a powerful organization known as the Teutonic Knights. This conflict began in the Middle Ages, and was representative of the com-plexity of the political situation in Poland then. Varmia, Copernicus's home after the death of his father, was situated between the lands of the Polish king to the west and those of the Teutonic Knights to the south and east.

The Teutonic Knights, a group which had been formed around 1190 in the Holy Lands by German crusaders, was modeled on the Knights Templar and Knights Hospitallers. They wore white mantles decorated with black crosses embroidered in gold. Although originally conceived as a Christian order—its members proclaimed the traditional vows of poverty, obedience, and chastity—the Teutonic Knights were most noted for their military prowess. They saw their central mission to be the destruction of the enemies of Christianity.

In 1226, the Teutonic Knights declared it their duty to convert the pagan inhabitants of Prussia to what they believed was the one true faith. Waging unrelenting war on the *Prusi*, or Prussians, the Teutonic Knights battled over the area for fifty years until, in 1283, the Teutonic Knights put down the last revolt and declared Prussia Christianized. By this time, the knights ruled the entire area between the Wisla River in the west and the Niemen River in the east. In 1308, the knights established Marlbork on the Wisla as their seat of government. They also encouraged immigration from Germany. The newcomers trickled in and slowly blended with the original inhabitants, who were converted to Christianity, mostly by the tip of a sword. This migration of Germanic knights and settlers was a direct threat to the power of the ethnic Poles who had earlier settled the area.

The advance of the Teutonic Knights was made easier by the succession of conflicts between various Polish groups vying for power. In the fourteenth century, most

KNIGHTLY ORDERS

Three principal orders of knights developed during the Crusades—the religious wars fought in the Middle East between Christian Europeans and Muslims over possession of the Holy Lands, from the eleventh through the thirteenth centuries—are the Knights Templar, the Knights Hospitallers, and the Teutonic Knights. In the beginning, the orders were charged with protecting Christian pilgrims traveling to Jerusalem and other cities in the contested Holy Lands. Within a short time, however, the orders became engaged in defending outposts and combating the Turks, Arabs, and other non-Christian peoples.

The Knights Templar were founded in 1118 when nine knights took the vows of chastity, poverty, and obedience. Ten years later, the group was officially recognized by the Church. In battle, members of this military order wore white tunics with a large red cross on their chests.

The Knights Hospitallers originated at the Hospital of St. John of Jerusalem, which was founded before the First Crusade. During the Second Crusade, the Hospitallers entered into battle, where they could be identified by their black tunics with large white crosses.

Originally a part of the Hospitallers, the Teutonic Knights were organized in 1190 to protect German pilgrims. In 1199, the pope recognized them as a separate order. The Teutonic order was never very active in Asia Minor and reached its greatest glory fighting in Prussia and other areas of Eastern Europe in the late middle and early modern era. Teutonic Knights wore white mantles with black crosses embroidered in gold.

of the country was united under King Casimir the Great, who established the capital at Cracow and, in 1364, founded the great University of Cracow. After Casimir's death in 1384, his crown went to his grandniece who soon married the Grand Duke Jagellon of Lithuania. This was the beginning of one of the most powerful dynasties in Polish history. The newlywed duke adopted the name King Ladislas and in 1410, using a combination of Polish and Lithuanian forces, defeated the Teutonic Knights decisively at the Battle of Tennenberg.

The defeat of the Teutonic Knights was a turning point in the long-standing struggle between the Poles and the Germans. The Teutonic Knights began to lose authority and control, and the lands slowly reverted to the Polish king. The town of Torun, where Copernicus

Being adopted by his uncle, Lucas Watzenrode, would help young Nicholas Copernicus *(pictured)* receive education and prominence. *(Artist unknown.)*

was born, had been founded by the Teutonic Knights as an outpost, and Varmia had originally been set aside as a separate diocese by the order. (A diocese is a religious term indicating a specific area in which all of the resident Christians are ruled over by the sitting bishop.) It was only in 1466, six years before Copernicus was born, that Varmia was ceded to the Polish king.

Before 1466, rulers of Varmia had been considered princes of the German Realm. Lucas Watzenrode, as bishop, paid his allegiance to both the Catholic pope and the Polish king. He had to coexist between two enemies of the same faith: the Polish king who wanted to take more land from the Teutonic Knights, and the knights who wanted to recapture the lands they had lost. It took considerable political and diplomatic skill to rule in such a volatile situation. Lucas had the necessary skill; as a result, he was the most important man in Varmia.

Lucas Watzenrode was so successful at holding off the designs of the Teutonic Knights that they called him "the Devil in human shape." They considered him to be their most formidable opponent. The bishop was not shy about expressing his sympathies for the Prussian people, whom he said had been ruthlessly crushed by the knights. It is a recorded fact that the knights prayed for the death of the bishop every day. Copernicus would adopt his uncle's ideas as well as his enemies, and would maintain his own lifelong hatred of the Teutonic Knights.

The death of Nicholas's father and his subsequent adoption by his uncle were probably the most pivotal events in the future astronomer's life. Uncle Lucas was generous with his money, and his prestige opened up opportunities for his nephew that might not have been available otherwise. Watzenrode also wielded almost total control over Nicholas's life. Andreas, his older brother, did not adjust well to the pressure, but Nicholas accepted the guidance of his beloved uncle. When Lucas decided it would be best if the brothers entered the church, since he could easily arrange for them to take positions as canons in his diocese, Nicholas did not put up any resistance. He would have to wait for a canon to die before a position became available, however, and that might take some time. In the meantime, it was necessary for him to attend school, as posting an uneducated man to a Church position would not be acceptable. So, at the age of eighteen, Nicholas and his brother, Andreas, journeyed south on the Wisla River, where they enrolled at the University of Cracow.

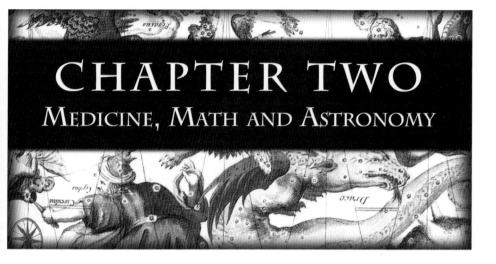

CHAPTER TWO
MEDICINE, MATH AND ASTRONOMY

One of the chief characteristics of the Renaissance is a renewal of the ancient Greek idea that men from the elite classes should receive a well-rounded education. King Casimir the Great founded the University of Cracow in 1364, and over time it became nearly as well respected as the best universities in Italy, France, and England. The main areas of study were philosophy, law, mathematics, medicine, astronomy, and literature, as well as Latin, Greek, Hebrew, and other modern and ancient languages. Copernicus knew several languages, including German, Latin, and Greek. He was especially proficient in Latin, which was the universal language for scholars during the Renaissance. The widespread use of Latin meant that

Nicholas and his brother attended the University of Cracow, located in the capital of Poland. *(Woodcut from Hartmann Schedel's* World Chronicle, *1493.)*

regardless of where a student was from he would be able to communicate with his fellow scholars in a common language. Latinizing one's name became a rite of passage for many learned men during this time. Even Nicholas would Latinize his name, from the Polish *Kopernik* to the Latin *Copernicus* we know him by today.

By the time Nicholas and Andreas arrived in 1491, the University of Cracow had developed a strong reputation in the fields of mathematics and astronomy, which were taught together. Mathematics at this time was a much different discipline than it is today. Then, it consisted primarily of arithmetic, algebra, and classical geometry. Today, the study of mathematics is viewed as essential to the understanding and advancement of science and technology. In Copernicus's time, while there were obvious uses for mathematics in commerce, construction, navigation, and a few other fields, the study of mathematics beyond this "practical" level was considered a purely intellectual pursuit, a sort of abstract game that could be studied for its own sake but had little value outside academia. The use of mathematics in science, which was in its early stages when Copernicus was a student, is one of the critical events in the development of the modern scientific method.

One of the mathematical skills Copernicus learned was how to position and calculate relative distances. He probably used a text written by Johann Mueller, better known as Regiomontanus, which was the Latin name of

During the Renaissance, the sons of the middle class, who made their living from trade and other commerce, began attending universities in greater numbers. *(Courtesy of Sächische Landesbibliothek.)*

his birthplace, Koenigsberg. Regiomontanus had continued the work started by George Peurbach of Vienna, a brilliant mathematician who died at a young age. Peurbach began developing a set of trigonometric tables that were highly accurate for their time. Regiomontanus continued to work on the tables and also on trigonometry, the study of triangles and their application. (The name "trigonometry" was not coined until 1613.) Regiomontanus gained fame as well for his *Ephemerides*, the first printed tabulations of planetary positions. He was also called to Rome in 1474 to supervise the reform of the old Julian calendar. When Columbus set sail across the Atlantic in 1492, he probably took an early version of Regiomontanus's *Kalendarium* to use as a navigational tool.

Much of the work done by astronomers during the Renaissance was concerned with the prediction of celestial events, such as eclipses, and tracking the positions of the planets within the zodiac. Astronomy was often combined with astrology, which postulates that the Sun, the Moon, and the planets influence the events in our lives after having determined our personality traits at the time of our birth. To make accurate predictions, an astrologer needed to know the configurations of the heavenly bodies at specific times. Astronomy was also used for navigation, agriculture, and calendar formation. Copernicus studied mathematics and astronomy at Cracow with a professor named Albert Szhamotuly, from whom he probably learned the signs of the zodiac, their positions in the sky, and the effects of locating the various planets in the different houses of the zodiac.

Copernicus differed from the typical astronomer of his era by taking a creative approach to the principles he learned. He was willing to rethink some of the basic assumptions if he thought doing so would better explain how the universe was structured. Surprisingly, his primary goal was to not to create a heliocentric model in order to "correct" the Earth-centered system. Copernicus merely wanted to restore something he thought had been lost to astronomy. Exactly what this was, and why it motivated him to create a new schematic for the planets, can best be seen in the context of the centuries of astronomical work that had come before publication of *On the Revolutions of the Heavenly Spheres.*

Human life is intimately connected to the solar system. In order to fulfill many of life's basic necessities, such as growing food and building shelter, we need an understanding of Earth's location relative to the Sun, the Moon, and the other celestial bodies, whose changing relationships determine the seasons, the tides, and other natural occurrences. This understanding was important while our ancestors were hunting and gathering, but became even more critical once people began congregating in towns and building civilizations. The ancient Egyptians, whose livelihoods depended on the annual flooding of the Nile River, developed a solar calendar based upon the appearance and disappearance of the star Sirius, which coincided with the river's rise.

The Babylonians, who lived in present-day Iraq, went beyond studying the stars in order to fulfill their practical needs, incorporating the belief that occurrences in the heavens were omens of future events. Babylonian astrologers needed to make extensive observations in order to discover patterns they could use to predict eclipses and other celestial events. They devoted considerable resources to observing the sky and cataloging what they saw. This effort, which stretched over centuries, provided the Babylonians with a substantial numerical record of astronomical events.

The Moon was the body most frequently observed by the Babylonians, which led to the development of a calendar based on its phases. The problem with lunar calendars, which are still in use today, is that a yearly calendar based on the twenty-nine-and-a-half-day phase of the Moon does not provide a uniform-length day, month, or year that also coincides with Earth's seasons.

To further complicate things, the Babylonians discovered when they compared the lunar phases to the solar year of approximately 365.25 days, they still could not create an adequate yearly calendar. Neither system was divisible into twelve even months with either twenty-nine, thirty, or thirty-one days. For this reason we now have a calendar year made up of months of differing lengths, as well as the system of leap years.

The Babylonians first attempted to solve this problem by inserting a thirteenth month when it was necessary to fill out the year, but this was obviously not a satisfactory long-term solution. Eventually, they drew on their database of observations and developed an ingenious device to systemize the relationship between the solar year and lunar month. The Metonic Cycle, first instituted in 380 B.C., was based on the fact that nineteen solar years (approximately 6,935 days) nearly equals 235 lunar months of 29.5 days (approximately 6,932.5 days.) They still needed an "intercalary," or inserted, month occasionally, but it could now be added in a regular pattern of seven months every nineteen years.

Another problem the Babylonians had to deal with was the uneven length of the seasons. The solar year of

At the university, Nicholas was taught the model of the universe developed by the ancient Greek philosopher Aristotle, which featured the Sun and all the planets revolving around a stationary Earth. This model was incorporated into the Christian world view. The illustration shows a theologian and an astronomer talking about Aristotle's heavenly spheres. *(From Petrus de Alliaco, Concordancia Astronomie cum Theologia, 1490.)*

approximately 365.25 days is not divisible by four, and although there are four distinct seasons, the seasons are not uniform in length. They also discovered that the Sun does not travel at a constant speed during what they perceived as its yearly orbit of Earth. For one half of the year the Sun's speed seemed to gradually increase; the other half of the year it gradually slowed down. Both the increase and the decrease took place at a constant rate.

This same phenomenon of varying speed was noted about the Moon and the "wandering stars," or planets. The five known planets presented an even more intractable problem. They traveled eastward, as measured against the background of the stars, for several months before stopping, moving backward (westward) for several days, stopping again, then proceeding on their eastward course. Trying to explain this retrograde motion became one of the most perplexing problems, and driving forces, of planetary astronomy.

The ancient Greeks inherited some of the astronomical developments of the Babylonians, including the signs of the zodiac, perhaps the Metonic Cycle, and the vast treasure trove of numerical data culled from centuries of astronomical observations. To this inheritance the Greeks added their own characteristic concerns and philosophical assumptions.

The Babylonians had relied on numerical calculations to predict where objects would be at specific times. The Greeks developed a way to lift stargazing to a higher level of abstraction by incorporating geometry, the mathematics of spatial relationships, into astronomy. The Greeks did not restrict themselves to tabulating the numerical relationships of events. Instead, they developed geometrical models—pictures—of how the Sun, the Moon, stars, planets, and Earth operated as a system.

One of the fundamental goals of Greek philosophy was to reveal the unity of nature. From the beginning, the Greeks required that for any planetary model to be viable it must operate as an interconnected system. Thus, incorporating the Babylonian data and any new observations into a reasonably predictive model of planetary motion was quite a challenge.

One philosopher, Pythagoras (569-475 B.C), who lived in a Greek colony in Italy, created a school of philosophy based on the idea that all of nature was structured by divine, mathematical ratios. Years later, the mathematician and astronomer Eudoxus developed a model of the heavenly bodies enclosed in a system of nested spheres. The most influential of all of the Greek philosophers to write on astronomy, though, was Aristotle (384-322 B.C.). He revised Eudoxus's model by adding additional spheres between the planets, which made the model more mechanically sound.

Aristotle's revision of Eudoxus's original idea became the dominant model of planetary motion for centuries. There were problems with it, though. The system of nested spheres did not explain why some planets seemed larger and brighter at certain times of the year. (It was not yet understood that planets do not produce light, as stars do, but only reflect light from the Sun.) Also, if planets traveled in circular orbits around Earth, their relative size and distance should remain constant.

Aristarchus of Samos (310-230 B.C.) attempted to solve this and other problems by suggesting that Earth orbits the Sun, which would explain the changes in the size of the other planets. However, this solution was rejected because of two assumptions based on the earlier model: that the universe is finite, and that the stars are located on the universe's outermost sphere. Those who rejected Aristarchus reasoned that if Earth orbited the Sun the stars would change in size and magnitude as we whirled around the heavens. Because no change in the stars could be seen, this argument did not win many adherents and his theory never took hold. Aristarchus suggested there was no visible change because, contrary to popular belief, the universe was extremely large, making Earth's orbit insignificant in comparison. Today we call the phenomenon of the apparent change in size or magnitude of a star as seen from Earth "stellar parallax." Copernicus would later use Aristarchus's argument to explain the lack of detectable stellar parallax.

After Aristarchus's heliocentric model was rejected, there was still the problem of how to explain irregular planetary orbits and retrogression in a unified model based on circular orbits. Today, because of the work of Johannes Kepler, we know that orbits are not circular; rather they are near-circular ellipses. But back then, even though no circular device could reproduce exactly

PARALLAX

The phenomenon of parallax is defined as the apparent shift in position of an object when viewed from different vantage points.

Parallax can be observed on a small scale by looking at an object (a clock, for instance) from across a room. Closing your right eye, hold a pencil at arm's length so that it blocks your view of the clock. Now, close your left eye and open your right eye. The apparent shift of the pencil represents parallax.

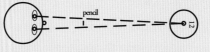

what astronomers observed in the night sky, a well-designed circular model came so close that astronomers found it intellectually difficult to give up on the idea.

Around 200 B.C., two new devices, the eccentric circle and the epicycle, were introduced in order to help more closely align geometrical models based on Aristotle's concentric spheres with the observable motion of the planets. The goal was to make a model that retained circular motion while accounting for the retrogression and apparent lack of uniform motion. By having the planets orbit on a circle that was eccentric (off-center) relative to the starry sphere, each planet still traveled at uniform speed on a circular orbit. But because the orbit was not centered in relation to the fixed stars, which were believed to whirl around Earth every twenty-four hours, the eccentric allowed a planet's velocity to vary as measured against the outer most sphere while still maintaining uniform circular motion.

The epicycle was a small circle centered on the larger orbit, called the deferent. A planet traveled on an epicycle, which in turn orbited on the deferent. This allowed the planet to have two uniform circular motions. For part of the epicycle rotation, the planet travels *with*—in the same direction as—the deferent. For the other period the epicycle travels *against* the deferent. This model explained retrograde motion as occurring when the planet on the epicycle traveled against the deferent. From Earth, the planet would appear to be moving backward.

These two geometrical devices explained the perceived retrogression and lack of uniform motion of the planets. Each planet's eccentric and deferent-epicycle system was unique and determined by numerical data gathered from observations. This system was not perfect, due to the fundamental problem that orbits are not perfectly circular. Still, the introduction of the eccentric and the epicycle made the concentric sphere model more efficient and useful for predictions.

One last device to be introduced in pre-Copernican astronomy came from the most creative and brilliant of all the ancient astronomers, Claudius Ptolemy, who lived in Alexandria, Egypt, during the second century A.D. Founded by Alexander the Great, Alexandria served as a center of the Greek-influenced, or Hellenistic, world. Later, after the Roman Empire collapsed and Europe entered the Middle Ages, Alexandria would be one of the

Opposite: Ptolemy and Regiomontanus shown on the frontispiece of the Epitome of the Almagest, Regiomontanus's translation of Ptolemy's ancient text. (Engraving by unknown artist.)

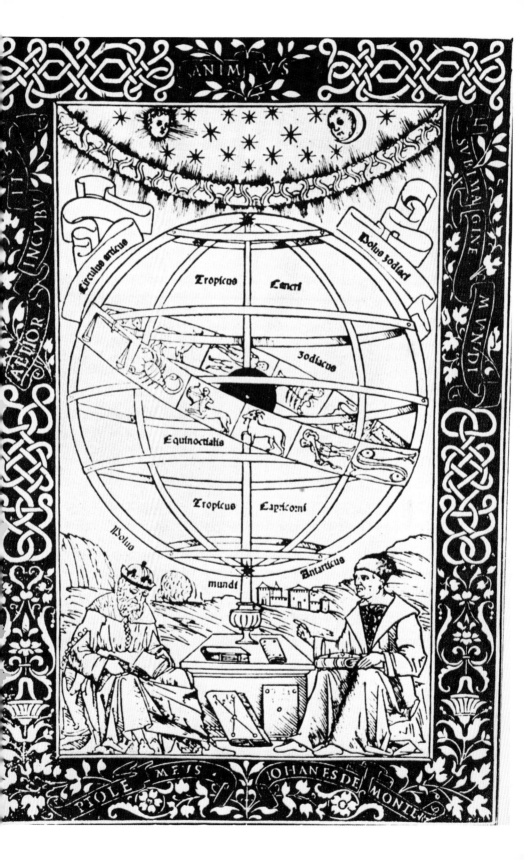

places where the achievements of the ancient Greeks were maintained and revised until their reintroduction in Europe during the twelfth and thirteenth centuries.

Ptolemy was the last great astronomer of the ancient world. Much of his work has been lost to time, and most of what survives is included in a work usually referred to as the *Almagest*, a name that comes from its Arabic title and means "The Greatest." The *Almagest* is the ultimate geocentric application of Greek geometry and Babylonian numerical data of the observed motion of the heavenly bodies. Largely a set of tables and functions that can be used to locate the planets, the Moon, and Sun at a specific time, it also contains a catalog of over one thousand stars grouped in constellations.

Ptolemy applied the eccentric and epicycle-deferent to his obversations, but still found he was unable to predict planetary positions to the degree of accuracy he sought. To compensate, he introduced another device— a point within the deferent called the equant. Ptolemy located his equant for each orbiting body at a point of equal distance on the opposite side of the center from Earth. The equant was located at the mirror image point to the Earth, opposite the center of the eccentric circle. This meant that a planet went faster when closer to Earth and slower when further away. Ptolemy was willing to reject uniform motion in order to make his model more useful and accurate in predicting planetary motion.

It is important to remember Ptolemy and his predecessors were primarily concerned with creating a model

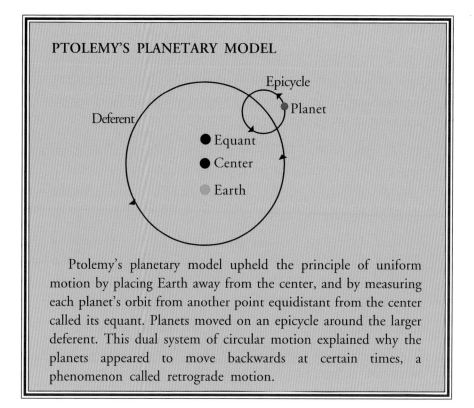

PTOLEMY'S PLANETARY MODEL

Ptolemy's planetary model upheld the principle of uniform motion by placing Earth away from the center, and by measuring each planet's orbit from another point equidistant from the center called its equant. Planets moved on an epicycle around the larger deferent. This dual system of circular motion explained why the planets appeared to move backwards at certain times, a phenomenon called retrograde motion.

that fit the numerical observations as much as possible. While it is unclear whether Ptolemy thought his model actually mirrored physical reality, it is probably safe to say this concern was secondary. His principal interest was to devise a unified geometrical model that could be used by anyone who understood the mathematical tables and functions.

Later, Copernicus was compelled to revise Ptolemy, but not simply because he wanted to correct this gap between physical reality and mathematical abstraction. Rather, he took issue with a central aspect of the model. Although we do not know exactly when, at some point

Ptolemy developed an elaborate model of the planetary system, based on the ideas of Aristotle, that included the use of deferents and epicycles. *(From Hevelius, Selenographia, 1647.)*

Copernicus became convinced that Ptolemy's equant violated Aristotelian principles. While motion was uniform as observed from the equant, it was non-uniform along the deferent. He was convinced that any accurate model must maintain both circular orbits and uniform motion as viewed along the deferent. Any compromise on this point was a violation of the principle of uniformity that had been the bedrock of Greek philosophy.

A central book in Copernicus's astronomical education was a medieval text called *The Sphere*. Written in the thirteenth century by Johannes de Sacrobosco, it was used throughout Europe as an astronomy text until the seventeenth century. There is also a good chance that Copernicus studied privately with a noted Polish astronomer and mathematician named Albert Brudzewski, although there is no record of him formally attending Brudzewski's lectures. Sometime during this period of study, Copernicus also learned how to make astronomical observations and to use the rudimentary observational instruments then available.

Aristotle's system and its influence included more than the planets. In his work *On Physics*, he divided the universe into two spheres. The first area, which included Earth and stretched to the Moon, was the sublunary region. The second sphere included the rest of the universe, from the Moon to the outer sphere of stars. Four basic elements—earth, air, fire, and water—were the building blocks of the sublunary region. Of these elements, earth was the heaviest, which explained why the planet Earth was considered centrally located. The four elements intermixed and transformed, but each element had its unique traits, or inherent motivations. Earth was constantly compelled to fall, for example; water, the next heaviest, was drawn to earth. Fire and air, the lightest elements, were ceaselessly attempting to escape upward. Aristotle saw the intermixture of the elements that comprise the sublunary sphere as motivating the changes our world is constantly going through.

All was different in the permanent, changeless heavens, which were made up of a fifth element, aether, an eternal substance, immutable over time and distance. Outside the sphere of the fixed stars existed the "prime mover," the primordial force that turned the spheres. The prime mover turned the first outer sphere, creating energy that then passed through that sphere to the next one. Each sphere's turning induced the next to turn, like interlocking gears. The energy of the universe came from the outside and moved inward toward Earth.

Aristotle's philosophy had become dominant by the fifteenth century. It was a philosophy, however, and was not based upon what we have come to see as the basic scientific techniques of experimentation and quantitative evaluation. The entire world system was based only on observation and logic. Aristotle rejected the use of experimentation, and the idea that mathematical analysis could both describe and reveal nature.

Another reason that more astronomers did not scrutinize the geocentric planetary system was the influence of religion on scientific thought at the time. The accepted model fit with several passages in the Holy Bible that mention the Sun moving and Earth remaining motionless. These passages in the Bible kept many believers from questioning the accepted Greek model. Combined with Aristotle's interpretation of the physical evidence, this served as a powerful roadblock against considering the creative idea of a rotating and revolving Earth.

Things began to slowly change in the Renaissance, when natural philosophers, the forerunners of the modern scientist, began to analyze the ancient texts and to formulate new opinions and philosophies based on these. While we do not know precisely when Copernicus began to question what he learned in school about the structure of the universe, we do know he received an education that would provide him with the insight and the skills needed to question conventional wisdom and to begin the process of rearranging the heavens.

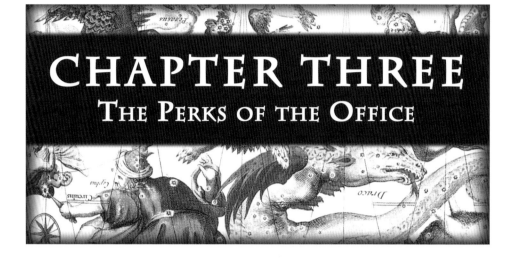

CHAPTER THREE
THE PERKS OF THE OFFICE

In 1494, twenty-two-year-old Copernicus was called away from his studies at the University of Cracow to travel to Lidzbark Castle, residence of the bishop of Varmia and the functioning seat of government. His uncle had sent word that it might be the right time for him to receive his appointment to the canonry at Frombork. However, the elderly canon of the Frombork Cathedral, Matthias de Launau, died on September 21, 1494, a few days too early for Copernicus to receive his post. In a peculiar arrangement, Uncle Lucas's power to appoint candidates for these positions was only in effect during the even-numbered months of the year. During the odd-numbered months, the power belonged to the pope. Canon Launau's death in the ninth month of the year made Copernicus's trip somewhat a waste of time. He returned to Torun for a brief period to see old friends and to conduct business.

He hoped to return to school and continue his studies and was relieved when his uncle agreed to fund more education. He had left the University of Cracow without attaining a bachelor's of arts degree, which was not uncommon. Only about a fifth of the students at Cracow ever received a certificate. A degree would be helpful, however, if Uncle Lucas got the chance to appoint his nephew to the canonry, which technically required a degree in either theology, medicine, civil (Roman) law, or canon (church) law. The decision was made that he would study canon law with the goal of receiving a doctorate. This was the same degree that Lucas Watzenrode had received the year Copernicus was born.

Canon law was the collected body of rules, regulations, and laws that governed the Catholic Church in both its day-to-day operations of its lands as well its negotiation of the more abstruse spiritual matters in the various religious communities and orders. It was just the kind of knowledge an influential churchman should have.

Instead of returning to Cracow to study, Copernicus was sent to the very heart of the Renaissance. In the fall of 1496, he crossed the Alps and arrived in the Italian city of Bologna. The University of Bologna was considered the best place in Europe to receive an education in

instructors lived in large houses and rented rooms to the most promising students, which made private instruction easier and also increased their income. Many students took advantage of these arrangements and learned a great deal from the close contact with their teachers.

In 1497, a vacancy opened up at Frombork Cathedral during an even month. Uncle Lucas submitted the name of his nephew, and Copernicus was elected. (Officially, these positions were chosen by election, but the chances of anyone crossing Uncle Lucas and not voting for his candidate were very slim.) Although the new job did not mean Copernicus had to return home right away, he could now begin receiving the canon's prebend (salary) in addition to the allowance he was given by his uncle.

While in Bologna, Copernicus studied privately with a famous astrologer, Professor Domenico Maria Novara. He also rented a room from Novara, which was no doubt appreciated by the professor, who let the young Pole collaborate on his work compiling astronomical data. In fact, Copernicus's first recorded observation was made while working with Novara. On March 9, 1497, it was recorded they observed the Moon approach Aldebaran, a star in the constellation Taurus. They watched as the Moon crossed in front of the more distant star, an "occultation" as it is called in astronomy.

In addition to working with Novara, Copernicus also studied Greek. Before learning Greek, he had been dependent on the Latin translations, which were usually made from Arabic translations of the original Greek editions. Many scholars of the period who only knew Latin had to depend on reading these translations of translations. This resulted in many misunderstandings and much incorrect information being handed down for generations.

While studying in Bologna, Copernicus became more familiar with the works of Pythagoras. Pythagoras's belief that numbers governed the structure and the operations of all of nature, including what takes place in the heavens, had a profound influence on the young student. This idea that mathematics was the necessary tool to understand the world became one of Copernicus's guiding concepts.

Although Nicholas was a serious student who was interested in learning all he could, his brother Andreas did not make the same impression when he joined him in Bologna in 1498. There is little known about Andreas's scholarly career. Uncle Lucas apparently focused most of his attention on Nicholas's advancement. Whether this is because Nicholas was a more serious student than his older brother is not known for sure.

In 1499, the brothers found themselves without enough money to make ends meet. They appealed to their uncle's secretary, who was passing through Bologna on his way to Rome. The secretary arranged a loan, eventually repaid by Bishop Watzenrode, which kept them solvent until they received their next allowance from their uncle and Nicholas's salary from the church.

The year 1500 was a special year for the church, and

for Copernicus. It was declared to be a Jubilee year. The pope issued a papal bull (proclamation) inviting all Christians to attend Easter festivities in the Holy City. Copernicus and Andreas traveled to Rome for the celebration, where they acted as representatives from Varmia to the pope. Over two hundred thousand Christians attended the Easter ceremony that spring. Copernicus stayed on in Rome for the rest of the year, where he lectured on mathematics and astronomy to the city's scholars and scientists. While in Rome, on November 6, 1500, he observed an eclipse of the Moon.

As the Jubilee year drew to a close, Copernicus made arrangements to leave Italy and return to Frombork to be officially installed as a canon. Andreas was also required to return. Nicholas had been receiving a canon's salary for four years, but had never actually set foot in Frombork. After making the treacherous journey north across the Alps, the brothers arrived in Varmia in the spring of 1501. On July 27, they attended the ceremonies at the Frombork Cathedral before the senior members and were formally admitted as canons in the Catholic Church.

CHAPTER FOUR
THE RETURN TO ITALY

The brothers' return to Varmia was to be short-lived. Once they were installed as canons, Nicholas and Andreas petitioned their uncle for an additional three-year leave of absence in order to continue their education. He granted them both permission to do so.

There were practical reasons for Copernicus to return to Italy. He had yet to take his final examination in canon law. However, Copernicus also appealed to his uncle that in addition to completing his law degree, he wanted to study medicine. This idea appealed to the bishop. Trained physicians were valuable, especially in remote places such as Varmia. This new goal resulted in the brothers parting ways. Andreas headed back to Rome, where he had made connections during the Jubilee year, and

In 1501, Copernicus moved to Padua to study medicine at the city's university.
(From Hartmann Schedel, World View, 1493.)

Copernicus was bound for the university at Padua, the principal college for the study of medicine.

Medicine might seem like a strange subject of study for an astronomer. At the time, however, medicine was more art and conjecture than it was science. One of the main diagnostic tools was astrology, because a person's astrological chart was thought to indicate his or her overall physical constitution. The movement of the planets not only governed aspects of character and destiny, it was believed each sign also affected a different portion of the body. For example, the sign of Aries governed the head, Leo the heart, Taurus the arms and shoulders. A chronic condition could be the result of the person's sign in the zodiac; a temporary illness could be the result of an unfavorable conjunction of planets in that house.

A typical treatment for an ill patient was to bleed them. A sharp instrument was used to open a vein and drain off a certain amount of blood. By doing so, it was believed that bad spirits, or "humours," were released. Often, the timing and the amount of bleeding were determined by astrological computations.

Another method of treating a patient was the use of medicines. Generally, these medicines were what we would call potions. A patient might be told to put "the worms with many feet that are found between the trunk and bark of trees" in some wine or water and drink it. Accompanying the dose, the patient was instructed to recite the Lord's Prayer several times. An actual recipe

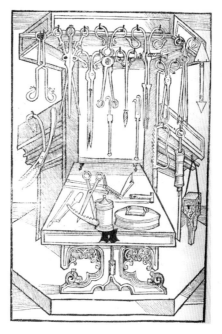

Some surgical instruments from the fifteenth century, including tongs, pincers, a saw, and scissors. *(Courtesy of the New York Academy of Medicine Library.)*

This fifteenth century illustration depicts a university professor lecturing students on the use of herbs in medicine. *(Courtesy of Sächsische Landesbibliothek.)*

for such a prescription was found handwritten on the back cover of Copernicus's copy of Euclid's *Elements of Geometry*: "Armenian sponge, cinnamon, cedar wood, bloodroot, dittany, red sandalwood, ivory shavings, crocus (or saffron), spodumene, camomile in vinegar, lemon rind, pearls, emerald, red jacinth and sapphires; a deer's heart-bone or pulped heart, a beetle, the horn of a unicorn, red coral, gold, silver and sugar." Other recipes included lizards boiled in olive oil, earthworms washed

This illustration (from the same volume as above) shows a group of university students listening to a professor lecture on medicine. *(Courtesy of Sächsische Landesbibliothek.)*

in wine, calf's gall, and donkey's urine. Medicine was often based on magic intermixed with religion. Some used peasant charms to ward off the evil spirits believed to cause illness. The application of these methods was also timed to coincide with the most advantageous astrological position possible.

Although much of medicine was still based on superstition, progress was made in some areas, such as the study of human anatomy. Leonardo da Vinci was one of the world's first anatomical artists. He made detailed drawings of the skeletal and muscular systems, and sketched the structural outlines and placement of the internal organs. The development of the craft of lens making led to a deeper understanding of how the eye worked, which then led to the development of ophthalmology. The study of gynecology and obstetrics was being formalized, and this helped to lower the death rate during childbirth.

Dissection was important to the early study of anatomy. *(Courtesy of the New York Academy of Medicine Library.)*

The changes that were taking place in education and in the very nature of learning were a result of the humanistic movement, which was at the heart of the Renaissance. Beginning in the middle of the fourteenth century, writers such as the Florentine poet Petrarch revived interest in the works of the ancient Greek and Roman writers. Although many of these works were studied during the Middle Ages, they were always interpreted from the Christian point of view. The humanistic scholars insisted that the works should be understood on their own terms. In other words, it was possible that the Greek philosopher Plato or the Roman poet Virgil had included ideas in their works that might not fit Christian theology. These scholars were not rejecting Christianity; they wanted to gain insight into nature and to approach human existence on its own terms. As the Renaissance progressed, scholars looked more closely at human existence and focused less on the necessity of

The Italian poet Petrarch is known for his translations of ancient Greek texts and for developing the sonnet, a poem of fourteen lines. His writings would influence English writers, such as William Shakespeare and Philip Sidney, during the Elizabethan period. *(Courtesy of the Bibliothèque Nationale, France.)*

eternal salvation. They became more secular, and strove for a higher level of objectivity when studying nature.

As a student in Padua, Copernicus was immersed in the very heart of the humanistic movement. For example, one astronomy teacher named Pomponatius, although a committed Aristotelian, taught his students not only to study the ancients but also to question their conclusions.

As it developed, humanism began to openly question what had always been viewed as "common knowledge." This necessarily led to the weakening of the bonds of social control, both political and religious, that had so tightly restricted society in the years since the collapse of the Roman Empire. Humanism was at least partly responsible for the fracture of the Catholic Church in the sixteenth century known as the Protestant Reformation; it is almost impossible to imagine the Reformation happening without the humanist movement preceding it. While some who became humanists deserted the church, others, such as Copernicus, remained Catholics while attempting to reform the church from the inside. Copernicus questioned authority in the area of science and the structure of the universe. He left it to others to advocate for religious and political liberalization.

Copernicus studied medicine at Padua until 1503. With only one year of his leave of absence from his duties in Frombork remaining, he still had not completed his law degree. He left Padua and traveled to the smaller city of Ferrara with the intention of completing the examinations for his doctorate of canon law. We do

Copernicus took his law exam at the university located in the city of Ferrara.
(*From Hartmann Schedel,* World View, *1493.*)

not know with certainty why Copernicus went to the smaller University of Ferrara to get his law degree, but the reason was probably financial. In Bologna, where he had studied law, graduates were expected to throw a huge party to celebrate the completion of their degree. A celebration on that scale could be quite costly. By removing himself to a university where he was not known, Copernicus could save himself from this obligation. He may also have been seeking a quieter place to study for the difficult law examination.

Copernicus received his degree in canon law in May of 1503 at the archbishop of Ferrara's palace. There is not record of what he did in the next few months. He might have lingered in Ferrara for a few months, where it is possible he read the work of the great Arab astronomer al-Farghani. A translation of al-Farghani's book *Elements of Astronomy,* a non-mathematical explanation of Ptolemy's system, had been published in Ferrara a few years before. It is known that Copernicus was familiar with the work, and given Copernicus's love of book buying, he could have purchased a copy during these months.

Copernicus could also have returned to Padua to once again take up his medical studies. No record of his ever taking a degree in medicine has been found, and it seems likely that he did not. He did, however, acquire enough knowledge to treat patients during his years in Varmia.

It was apparently during this time that his brother, Andreas, contracted the unknown disease that would prematurely end his life. Upon returning to Varmia, his behavior ruined what remained of his reputation with his colleagues in the church. Eventually, his illness caused him to deteriorate to the point where, in 1508, he requested and was granted another leave of absence to seek medical care in Italy. He returned to Frombork by 1512, but by then his appearance was so hideous, his body and face wracked by either leprosy or the lesions that accompany syphilis, that he was stripped of his position and his income as canon. He was even forced to repay an advance of twelve hundred gold florins and was told to leave the area. Andreas fought back. He stayed in Frombork, appealed to higher church authori-

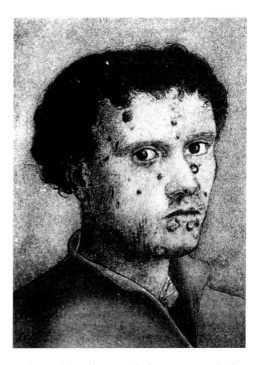

A young man afflicted with syphilis and the characteristic pustules of the disease, 1523.
(Courtesy of the Fogg Art Museum, Cambridge, Massachusetts.)

ties, and continued to show his diseased face around the town. Finally, the other canons, including Nicholas, agreed to reinstate his salary if he would agree to leave. Andreas left Poland forever and returned to Italy, where he eventually died. The exact date of his death is unknown but, given the advanced stage of his disease when he left Poland, he was probably dead by 1520.

CHAPTER FIVE
CHURCH AND STATE

On the way home from his medical studies in Padua, Copernicus stopped for a short visit in Cracow. He visited old friends from his university days, and also relatives, notably his sister and her merchant husband. He spent some time with his former professors and discussed some of the ideas about the arrangement of the universe he had developed over the last years. He was probably beginning to formalize his thinking about the Sun-centered planetary system. He would have most likely preferred to stay in Cracow, teach at the university, and continue thinking about and researching his astronomical theories. He was comfortable in the university surroundings, having never really done anything else since he was eighteen. He was now thirty and one of the

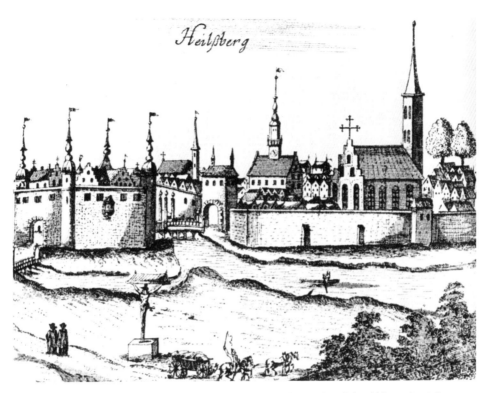

In one of his first official duties as a canon, Copernicus joined his uncle at the Lidzbark castle as the bishop's personal physician. *(Courtesy of the New York Public Library.)*

most educated men in Poland. He had pledged to return to Varmia to work in the church with Uncle Lucas though, and it was now time to make good on his promise.

In addition to being his uncle, Lucas Watzenrode was Copernicus's superior in the church hierarchy. The years Lucas had spent grooming Copernicus and paying for his education were intended to train his nephew to help in the administration of the diocese. When Nicholas returned, Lucas was glad to have someone around whom he could trust implicitly. His was a political position; it took every ounce of his savvy to maneuver between the Polish king, the pope in Rome, and the Teutonic Knights on the frontier. Lucas knew Copernicus could be relied on, and soon after his nephew returned from Cracow, he reported to duty with his uncle at the castle at Lidzbark.

Located several miles south and east of Frombork on the Lyna River, Lidzbark consisted of a castle with a wall and moat surrounded by a small village. It was very much like Frombork. Defensive fortifications were still necessary for all towns and cities due to the ongoing hostilities between the Poles and the Teutonic Knights.

A fair amount of the church's wealth had been spent in furnishing the residence of the bishop of Varmia. Copernicus discovered that he would spend his next years living in luxury, a drastic change from the relatively informal life he had been living as a student. The decor in the Lidzbark castle was very comfortable and meals were lavish. At mealtime, the bishop would appear with mitre, staff, and purple gloves to lead a procession into the Hall of Knights. There, a raised dais separated the tables so that diners were segregated according to rank. The table at the top of the dais was for the bishop and his guests. The next table down would seat the higher officials and was probably where Copernicus took his meals. The third table was for the lower officials, and the fourth through the eighth tables were for the servants and any common visitors, also segregated according to social rank. Jugglers, jesters, and clowns who entertained the assemblage occupied the ninth table.

Officially, Copernicus's job title was the bishop's personal physician. He also attended the sick at his uncle's court. However, his duties were quite wide-ranging. He also drafted most of the letters and documents necessary for the bishop to conduct government business, recorded the proceedings of official meetings, and kept track of treaties and trade agreements. He served as a traveling companion to the bishop, looked after the bishop's health and, particularly after gaining some experience, was his principal advisor and confidant. As part of his duties as bishop of Varmia, Lucas Watzenrode often had to attend the diets, or formal meetings, of the local leadership, who were electors of the Holy Roman Empire and were responsible for its governance. Often, these meetings of the local Polish diets conflicted with the bishop's schedule; in those cases, Copernicus frequently attended as his uncle's delegate with full authority to speak for the bishop.

Many came to believe that Copernicus was his uncle's choice to succeed him as bishop. Certainly, Lucas Watzenrode had spent years grooming Copernicus. It is possible, though, that the bishop came to realize that, although his nephew had the intellectual skills to rule and was honest and fair, he lacked the personality to succeed at the job. The bishop had to make some ruthless decisions in the pursuit of his lifelong goal of forcing the Teutonic Knights from the Baltic coast, Prussia, and Pomerania. At one point, the bishop almost succeeded in having the entire Teutonic Order transferred to

Wallachia, in what is now Romania, to fight the Muslim Ottoman Turks. Besides the knights, there were several other political forces within Varmia and Poland that would have liked to see Lucas Watzenrode eliminated. His was a position that came with enemies as well as friends and sometimes it was hard to tell them apart. It is difficult to imagine quiet, scholarly Copernicus succeeding in such a public role.

In 1506, a new king, Sigismund I, sometimes called "the old," assumed the Polish throne. Soon after taking the throne, Sigismund I ordered Grand Master Albert,

King Sigismund I of Poland came to the throne shortly after Copernicus moved to Lidzbark. *(Portrait by unknown artist. Courtesy of Wawel Royal Castle, Cracow, Poland.)*

the leader of the Teutonic Knights, to pay him allegiance. This would consist of Albert meeting with the king and declaring himself a vassal during a public ceremony, as well as making a substantial financial contribution from the knights, which Albert would have to raise from the coffers of his own lords. At this time in history, the region was under the feudal system, in which each lord of a manor owed his allegiance to another, higher placed lord, just as his vassals, or peasants, owed him allegiance for being allowed to live on his land. The land was called a fiefdom after the word "fief," which entailed the promise of service from the subordinate manor lord to the next higher ruling power. The domain of the knights had been a fief of Poland since their defeat at Tannenberg in 1410, and Sigismund was within his rights to demand Albert's allegiance. The grand master of the Teutonic Knights refused the king's request, however, and Sigismund became furious, nearly going to war over the insult. Although the bishop was delighted to see the king so angry with his enemy, the prospect of armed conflict was not a good one for Varmia, which could easily be absorbed into Poland during a war. Fortunately, cooler heads prevailed and war was avoided, at least temporarily.

Although busy working on matters of church and state, Copernicus found time to pursue his intellectual interests. One of the traditions of the time was for scholars to translate texts from Greek into Latin so that more scholars would have access to classical thought.

Grand Master Albert, the leader of the Teutonic Knights, converted to Lutheranism after assuming control of Prussia.

Humanism, the dominant intellectual ideology of the time, naturally had opponents who objected to the high pedestal on which ancient learning, especially Greek learning, was placed. Most anti-humanists worried that focusing on classical, myth-based literature would lead Christians into paganism.

In 1509, Copernicus ignored this criticism and found a volume from antiquity to translate. Instead of choosing an ancient text, however, Copernicus chose to translate a collection of seventh century Christian epistles written in Greek by Theophylactus Simocatta, a Byzantine historian whose best-known work was an account of the reign of Emperor Mauritius. Simocatta's *Epistles* were divided into three categories—pastoral, amorous, and moral—and are generally considered by scholars to be dull and overly pious. Even though it was a Greek text, its content was not as valued as the works of the ancients. This is a characteristically cautious move by Copernicus. He wanted to maintain his reputation as a scholar but, at the same time, to avoid entering the controversy over humanism. His choice of a Greek text by a Christian author would not conflict with those who opposed the revival of pre-Christian Greek learning. The book's publication attracted little attention.

One interesting aspect of the *Epistles* is the inclusion of an introductory poem with the collection. Apparently, Copernicus asked a friend named Rabe (meaning "raven") to compose some verses for the occasion. Having Latinized his name, Rabe was formerly known as Laurentius Corvinus, which also meant "raven." Rabe worked in the city of Breslau as a town clerk. The verses began by praising Copernicus's loyalty to his uncle, who was his patron, and went on to praise the classical tradition. Rabe's poem also talks about how Copernicus "explores the rapid course of the moon and the changing movements of the fraternal star and . . . with the planets . . . he knows how to explore hidden causes of things." This indicates Copernicus had discussed his astronomical theories with his friend, and that they were so important to him, and so impressive to his friend, that they warranted mention in this poem.

In the spring of 1512, the bishop traveled to Cracow to attend the wedding of Sigismund I and the coronation of the new queen. Copernicus, as usual, traveled with his uncle, but did not return with him. During his return to Lidzbark, Bishop Watzenrode suddenly became ill. By the time he reached the castle, the bishop had a very high fever. He died three days later, on March 29, 1512. Many suspected the Teutonic Knights poisoned him. There is no record of Copernicus's reaction to his uncle's death, and it is not known if he blamed himself for not being with his uncle as he usually was. There was probably little he could have done, though, as the bishop was in his mid-sixties and had led a hard and stressful life.

His uncle may have groomed Copernicus to become the next bishop, but it seems that such a promotion was not seriously considered by high church officials.

OPPOSITION

A planet is in opposition when it is opposite the Sun as seen from Earth. The best time to observe the planets is when they are in opposition, as this is when they are generally closest to Earth. The opposition effect is most easily noticed with Mars.

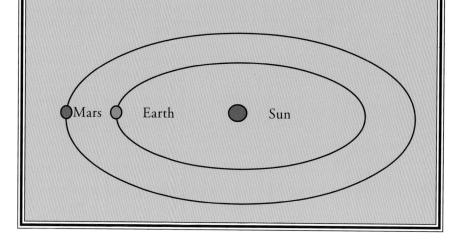

Copernicus left Lidzbark soon after his uncle's death to assume his duties as canon at Frombork. At the age of thirty-nine, fifteen years after being appointed to that position, he was finally taking residence at the cathedral castle in the north of Varmia. There he would turn out to have unburdened hours to pursue his primary passion, astronomy. Records show that Copernicus was doing astronomical work from his new home as early as June 5, 1512, when he observed Mars in opposition to the sun.

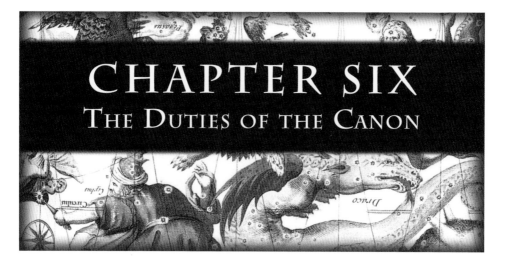

CHAPTER SIX
THE DUTIES OF THE CANON

The sixteen canons who lived in Frombork ruled over the church's lands the same way a nobleman controlled his manor. As noblemen were always subject to the king, so the canons were subject to the bishop, the Polish king, the Holy Roman emperor, and ultimately the pope, in the complex chain of duty and allegiance that was the feudal system. The canon's powers and responsibilities extended to secular affairs. They managed the local commerce, supervised foreign trade, oversaw court proceedings, and meted out justice for those convicted of crimes. Canons collected taxes, rents, and tithes, and appointed mayors and other town officials. They were also responsible for the defense of their land and people.

Only one of these sixteen canons had taken the higher

Copernicus was a canon at Frombork during the last years of his life.
(From a 1684 copperplate in Jan K. Hartknoch's Prussia: Old and New.)

vows and was allowed to officiate at Mass. The others assisted in morning and evening services if they had not been called away on business, as was frequently the case. One of the canons during Copernicus's tenure worked as a secretary for the Polish king, while another served in Spain as ambassador. Copernicus himself had spent fifteen years pursuing his education and working for his uncle. After settling in Frombork, though, Copernicus devoted himself to serving in a variety of official capacities for the chapter and never again traveled abroad.

In 1514, Copernicus was invited to send his opinion to the Lateran Council on calendar reform. The invitation came from Canon Scultetus, Uncle Lucas's former secretary, who had arranged the loan for Copernicus and his brother in 1499, and who was now domestic chaplain for Pope Leo X. Copernicus declined the invitation with the comment that the council would be a waste of time because the motions of the planets, the Sun, and the Moon were not understood well enough to calculate a more accurate calendar.

During this period, the turmoil in the Catholic Church came to a head. In 1517, Martin Luther, a thirty-three-year-old German monk, posted ninety-five theses, or complaints, against the Church, on the door of the castle church in Wittenberg. His protest ushered in the Protestant Reformation, which in a short time would divide Christian Europe and lead to periods of religious wars.

MARTIN LUTHER was born in a small town in Saxony, a German principality, to well-to-do parents in 1483. Luther was a highly intelligent boy who at first planned to be a lawyer but later felt drawn to enter the priesthood. He joined the religious order of Augustinian monks. Unlike some others who entered the church, Luther was not motivated by career as much as by a deep-seated feeling of his unworthiness before God. Legend has it that when in fear for his life during a thunderstorm, he promised God he would become a monk if his life were spared. Upon becoming a priest, he studied theology at the University of Wittenberg, where he later taught and preached.

Because of his feelings of unworthiness, Luther diligently followed the church's suggestion and sought to alleviate his suffering through the doing of good works. Finding little relief, Luther became convinced during his twenties that it was through God's grace alone that one was saved. To think a person could save oneself through his own actions was to deny God's mercy. He struggled with these ideas for years.

Meanwhile, Luther grew increasingly disgusted with the corruption, such as the selling of indulgences, within the Catholic Church. (An indulgence was an official forgiveness for a sin.) Originally granted to the sinner after the completion of some "good work," the policy of exchanging indulgences for cash contributions to the Church had developed during the Middle Ages. In 1517, a new German archbishop began to actively push the selling of indulgences in order to pay off a loan. This act culminated with a climax in Luther's personal doubts about Catholic theology. On October 31, 1517, Luther nailed ninety-five theses, or complaints, on the door of the castle church, which served as a community-wide bulletin board. Most of the complaints were aimed at abuses in the Catholic Church. This was a brave, even rash, act, but it was not unprecedented. What lifted Luther's complaints above the run-of-the-mill was that in the process of attacking corruption he also criticized the Church's doctrine.

Luther originally intended for the theses to spark debate within religious circles, and he circulated them among other bishops and friends. The issue might have remained local, and Luther might have remained in the Catholic Church, were it not for the advent of the printing press, which allowed for wide distribution of Luther's theses.

Although many were sympathetic to Luther's criticisms, and some bishops wanted to find a way to keep him in the Church, the Papal Court labeled him a heretic and excommunicated him in 1520. Luther burned the Papal Bull of excommunication in protest, irrevocably severing his ties with the pope.

In 1521, Holy Roman Emperor Charles V asked Luther to attend the Imperial Diet of Worms. The emperor and the church leaders hoped Luther would use the visit as an opportunity to recant his teachings. The emperor granted Luther a pass of safe conduct to travel unhindered in the Holy Roman Empire for twenty days. Although branded a heretic, he was enthusiastically received on his journey. But when he refused to denounce his writings, Charles V decreed he could be killed as an outlaw, and the assassin would receive no punishment.

Luther posed a secular as well as a religious threat to the established order because many German princes wanted to slip from beneath the control of the Vatican in Rome. The pope demanded an allegiance, which included taxes, from secular rulers. When a prince aligned with Luther's movement, he no longer had to show allegiance to the pope and he could also confiscate the Church's property in his lands.

Upon learning there was a price on Luther's head, one of his patrons, Friederich the Wise, hid Luther at a castle in Wartburg for his own protection. While in hiding, he began the translation of the Bible into German. In 1522, Luther returned to Wittenberg, which had since become the heart of the Protestant Reformation. He spent the next several years preaching, writing, and organizing the Reformation.

In 1525, Luther married a former nun named Katharina von Borah, and they eventually had six children. For the rest of his life, Luther continued to preach his faith, and to attack the Catholic Church, despite increasing infirmities as he aged. In 1546, at the age of sixty-two, he was called to Eisleben, his place of birth, to negotiate a dispute between nobles. Upon the successful conclusion of the negotiations, he fell ill and died.

This illustration depicts Luther receiving the direct word of God, posting his ninety-five theses, and the conflict that ensued.

One year before Luther's actions, Copernicus was sent to take up duties as the new administrator for Olsztyn and Pieniezno, two districts in the south of Varmia some fifty miles from Frombork. For the next three years, Copernicus resided at Olsztyn Castle, on the Lyna River, along with an entourage of servants and assistants. There, he oversaw the daily operation of the two districts. His job consisted primarily of appointing laborers, officiating at the transfer of property between peasants, and signing documents in the presence of witnesses. All of these transactions were recorded in a daily journal he kept during his time at Olsztyn.

In 1519, Copernicus returned to Frombork just as war broke out between the Polish King Sigismund and the Teutonic Knights. Grand Master Albert had hired mercenaries to invade, loot, and pillage Varmia's peasantry, and Sigismund responded by attacking. At this point, Sigismund had to divide his forces to repel an invasion by the Tartars, a nomadic people from East Asia. To complicate matters, the Holy Roman Emperor Maximilian I died, which meant Sigismund had to attend to his duties as one of the electors responsible for choosing the successor to Maximilian.

After Sigismund drove the Tartars out, and a new emperor had been chosen, the Polish king focused on dealing with the knights. He again ordered the grand master to appear in Torun to pay homage. When the leader of the Teutonic Knights refused, as Sigismund knew he would, the king ordered the Polish troops to invade the Teutonic lands. This sparked a full-scale war.

For fifteen months, Varmia was the main battleground as the Teutonic Knights and the Poles engaged in fierce combat. The local civilian population was devastated; untold numbers of peasants—men, women, and children—were assaulted or killed by soldiers on both sides of the conflict. Houses and barns were burned; livestock was stolen or slaughtered.

When Grand Master Albert sent word that he was willing to discuss peace in the newly captured city of Braniewo, he asked Bishop von Lossainen, the man who

HOLY ROMAN EMPIRE

The Holy Roman Empire was founded on Christmas Day 800 when Charlemagne (Charles the Great), traveled to Rome to be crowned as the king of all of Christendom by Pope Leo III. By accepting the crown from the pope, Charlemagne showed his allegiance to the church and the pope acknowledged Charlemagne's secular rule over Christian Europe. Charlemagne had forged together many of the Germanic tribes that had destroyed the old Roman Empire and he now wanted to reorganize the old empire to its former glory. After his death the Carolingian empire, as it was called, eroded. It was not until the coronation of the German king Otto I in 962 that the title of Holy Roman emperor became an official designation.

The territory included in the Holy Roman Empire varied over the centuries, but at times it included most of present day Germany, Belgium, Luxembourg, Poland, the Netherlands, most of Switzerland, and sections of northern Italy.

There was a two-step procedure to becoming the Holy Roman emperor. First, a candidate was elected by an assembly of German princes called the Imperial Diet. Then the German king had to receive official coronation from the pope. In effect, the pope had veto power over the choice of the German princes, and this dual selection process inevitably led to tension between the secular king and the pope. Eventually, the German king was declared to be the king of the Romans by the electors, even if the pope had not crowned him Holy Roman emperor. In 1338, the electors claimed the right to elect the emperor, although emperors continued to be crowned by the pope until 1530. After 1438, the imperial office of Holy Roman emperor was controlled by the Hapsburg family.

The emperor was proclaimed to be the secular ruler of all of Christendom and was supposed to work in tandem with the pope for the good of all Christian Europe. In reality, the conflicts between different emperors and popes became one of the defining characteristics of political life throughout the Middle Ages and into the modern era. Several states never offered more than nominal allegiance to the emperor, whose power was always limited even in Germany. He, for example, could not involve himself in the internal affairs of a member state, and was even limited in how far he could act diplomatically without approval of the Imperial Diet.

After assuming dynastic control, the Hapsburgs made efforts to centralize power. After suffering defeat in the Thirty Years' War, which ended with the Treaty of Westphalia in 1648, these efforts gradually ceased. Caught between the religious splintering of Europe brought on by the Protestant Reformation and the rise in power of member states, such as Prussia, the empire declined. On March 6, 1806, Francis II, the last Holy Roman emperor, abdicated the throne and announced the Holy Roman Empire dissolved.

had succeeded Lucas Watzenrode, to represent Varmia in the negotiations. However, the bishop was either too sick or too afraid of treachery to go himself, and he sent two representatives instead. Copernicus was one of the negotiators. A document providing Copernicus with safe passage still exists. Written by Grand Master Albert, it states, "The honored highly learned Mr. Niklas Koppernick" should be given "free, safe, and Christian conduct, to and from" Braniewo. In addition, safe passage was also to be given to "his horses and servants."

The peace talks broke down, though, when Albert insisted on a promise of allegiance from the Varmia delegation. Copernicus and the others refused to commit to this condition. They left Braniewo without an agreement but glad to get away with their lives.

Early in 1520, the Teutonic Knights overran Frombork and set fire to the town. The cathedral was safe within its moat and behind its walls, but the knights laid siege to it for several months. Although most of the canons fled to the safety of Gdansk, Copernicus remained in the cathedral. He even spent some of his idle time during the long siege taking astronomical measurements.

When Sigismund signed a peace treaty with the Russians, with whom he had also been at war, he freed up more troops to use against the Teutonic Knights. Once again, Albert sued for peace, but he did not appear to want peace any more this time than he had before. He received some reinforcements from Germany and started up hostilities again.

At some point during the fall of 1520, Copernicus returned to Olsztyn to once again take up the duties of administrator. Because Olsztyn was far to the south and closer to territory controlled by the knights, much of the district had been laid to waste. Widespread murder and destruction had taken place there, and entire towns and villages had been burned to the ground. At this time, Copernicus and Heinrich Snellenberg were the only two remaining canons in the entire region of Varmia and they stayed together at Olsztyn Castle. The Teutonic Knights eventually laid siege to Olsztyn. Aside from a handful of servants, Copernicus and Snellenberg were the only defenders of the castle. Fortunately, the knights were tired of the war and the defenders were able to hold on and keep Olsztyn from falling into their hands.

In 1521, an armistice was signed at the encouragement of the Holy Roman emperor. Because of his brave and steady efforts during the war, Copernicus was made commissar of Varmia. This position gave him the authority to oversee the rebuilding of all the destroyed areas. The scattered and frightened peasants had to be found and resettled on their lands, a process that often caused fierce property disputes that Copernicus would have to mediate. Homes, barns, even entire villages and towns, had to be rebuilt. There was also a critical food shortage because no crops could be grown during the war.

In addition to rebuilding Varmia, Copernicus was given the task of composing a list of grievances against the Teutonic Knights in order to seek compensation for the damage they had done. His good friend and fellow canon Tiedemann Giese assisted him. They composed a document enumerating the burning of towns and villages, the pillaging of farms, the slaughtering of innocent people, and all the other horrific actions that take place when men go to war. This document was presented to the Prussian Diet, or assembly, when it met in the city of Grudziadz. The diet was responsible for making a final judgement.

Along with the list of grievances, Copernicus also presented his treatise on money. During the war years, he had seen the various coinages issued by the Poles, the Prussians, the Teutonic Knights, and even some of the major cities, become debased—less valuable—each year as less and less silver and gold was minted into the coins.

Tiedemann Giese, who became Copernicus's closest friend during his years at Frombork, encouraged him to publish his planetary theory.

Eventually, the money was almost worthless. Copernicus proposed to the Prussian Diet that all of the shoddy coins be called in and a new centralized authority, agreed to by all of the interested parties, be set up to issue new coins. Unfortunately, many of the nobles did not want to address the inflation issue because they had a personal interest in seeing inflation continue. Copernicus presented his monetary treatise again six years later, but the authorities still would not adopt his recommendations.

For several years following the armistice of 1521, negotiations took place between the Poles and the Teutonic Knights. Finally, a lasting peace treaty was enacted in 1525. Prussia became a fief of the Polish king and Grand Master Albert became a duke. He even converted to Protestantism after consulting with Martin Luther. After this peace treaty, Varmia regained all of the towns and villages occupied over the years by the Teutonic Knights.

After his duties in Olsztyn were finished, Copernicus returned to Frombork. He had been through war and seen much suffering. As he entered his fifties, many colleagues and friends were beginning to pass away, leaving him more isolated every year. He continued to take his turn as chapter envoy, which required him to tour the outlying estates, inspect how the local government was run, and collect the rents owed the cathedral. Often his good friend Tiedemann Giese, who would later become a bishop, traveled with him.

Copernicus was a good administrator who was always willing to offer suggestions to improve his homeland. For example, he instituted a plan to establish a standard size and price for a loaf of bread, a measure made necessary because of the inflation and rapidly fluctuating grain prices in the aftermath of war. He also helped draw up several of the new craft-guild regulations that were developing as more people left the countryside and began careers in the towns as skilled workers, apprentices, and journeymen.

Copernicus fulfilled his obligations as canon to the best of his considerable abilities. He was well known and respected. By any measure, his life was already a success. He still had one more contribution to make before he died, though. It would be a great work, and one that would require much of his time and attention as he grew older.

CHAPTER SEVEN
THE ASTRONOMER EMERGES

Copernicus had not been neglecting astronomy since returning from Italy. Sometime between 1509 and 1513, he circulated a small handwritten pamphlet called *Commentariolus*, or *Little Commentary*. In this manuscript, he communicated his heliocentric thesis to others for the first time. Many scholars in those days would circulate manuscript copies of their work to their colleagues to try out new ideas before committing to the expense of publication and the possible ridicule of detractors. Copernicus's *Little Commentary* helped establish his reputation as an astronomer with a bold idea. It also sparked an interest in his work that slowly spread throughout the educated elite of Europe.

The *Little Commentary*, which was rediscovered in the nineteenth century with the subtitle *A brief outline of Nicolai Copernicus's hypotheses on the heavenly motions*, proposed the radical thesis that the Sun is the central, stationary body in the planetary system, and Earth, along with the other planets, orbits the Sun. It also proposed Earth spins on its own slightly tilted axis once a day, which explained the apparent rising and setting of the Sun, the Moon, and the stars, as well as the changing of the seasons. Copernicus also commented

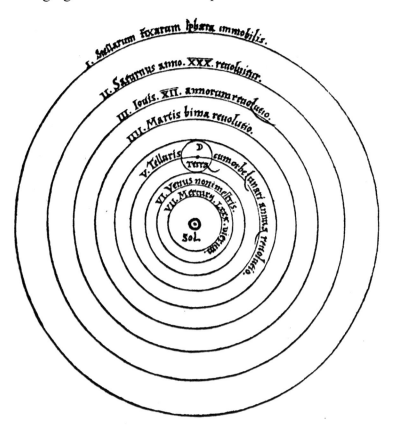

A diagram of the Copernican model of the universe, from his major work, *On Revolutions.*

on the need to revisit Ptolemy's model of the heavens because of its use of the equant, which Copernicus believed was too great a deviation from Aristotle. Copernicus claimed to have solved the problem of maintaining uniform circular motion by relocating Earth and Sun. He had created a model such that "everything would move about its proper center, as the rule of absolute motion requires."

In *Little Commentary*, Copernicus listed seven principles, or axioms, he said provided a more plausible explanation of how the planets, the Moon, the Sun, and the stars were organized. After listing out these axioms, Copernicus described the new circles, and epicycles— which were different from those devised by Ptolemy— of the Sun, the Moon, and the planets without going into mathematical detail or providing other observational evidence. He states he was "reserving these for my larger work." He concluded the treatise with a short paragraph: "Then Mercury runs on seven circles in all; Venus on five; the earth on three; and round it the moon on four; finally, Mars, Jupiter, and Saturn on five each. Altogether, therefore, thirty-four circles suffice to explain the entire structure of the universe and the entire ballet of the planets."

Although we do not know who received copies of the *Little Commentary*, it is logical to assume that among those to whom he would have sent his treatise were former professors and colleagues at the various universities where he studied, as well as individual amateur and professional scholars he knew personally in Cracow and maybe the universities in Italy and Germany.

Having propounded his theories in his treatise, Copernicus set about to show how they could be used to predict the positions of the Sun, the Moon, and planets. Fortunately, being a canon gave him time for his own work and a decent location for stargazing. Nicholas Copernicus's position in Frombork entitled him to an apartment within the cathedral walls, as well as two small estates in the countryside. His primary residence was in the tower that made up part of the cathedral's defensive fortifications. The tower was about fifty feet high and thirty feet square and was located at the northwest corner. There was a cellar, ground-floor kitchen, and maid's room in the lower sections, and a small living room and bedroom on the next level. A short flight of steps led from the second level to a third level containing a storage area and toilet. He had converted the fourth floor into a large workroom illuminated by nine windows, with an open air gallery at one end.

When a canon died at Frombork, the next most senior canon could move into the vacant residence. Although the better quarters were considered a benefit of seniority, Copernicus did not change residences as the older members departed because the tower offered him the clearest view of the sky.

Copernicus made many of his observations from the open gallery at the end of his workroom. However, a portion of his view was blocked by the workroom itself.

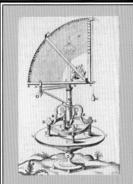

QUADRANT

The quadrant consists of a quarter circle, a moveable arm, and a plumb line. It was used to measure angles between celestial objects as well as the altitudes of stars. The curve of the quadrant is graduated with ninety degrees and positioned so its plane passes through the celestial object in question. The plumb line guarantees one side of the quadrant is horizontal, and the other is directed toward the zenith, or the point in space that is directly above the viewer. The moveable arm is used to site the object.

The quadrant can be traced as far back as Ptolemy (100-170 A.D.), and was used extensively by Islamic astronomers. Quadrants can also be used to tell the time of day. They were used even after the discovery of the telescope, because early telescopes could not measure the exact location of a star.

Medieval castles were built on the highest available spot for reasons of defense, which meant that Copernicus's tower was located at the highest point of the hill. This gave him a good vantage point from which to make his observations.

Although Copernicus had an ideal view of the sky in terms of his immediate physical surroundings, his geographic location was less than perfect. Frombork is so far to the north that it was sometimes difficult to get a clear view of Mercury, which, along with Mars, was one of the two main planets that interested him. (These two planets had the most eccentric of all the known orbits.) In addition, the proximity of the Baltic Sea and the Zalew Wislany, a freshwater lagoon between Frombork and the ocean, produced long periods of heavy fog.

Today, we think of astronomers working with telescopes, but they had not yet been invented in Copernicus's time. Galileo would be the first astronomer to use a telescope, in 1609. Copernicus owned a variety of devices that he used to examine the heavens, and he might have even constructed a few himself. He used a quadrant for observing the Sun, as well as an astrolabe for the stars. An upright sundial, its base set to point north and south, indicated the Sun's altitude at midday. He had an ancient triquetrum, an instrument used since Ptolemy's time for measuring the zenith distances of the planets and stars. In addition, he would have employed a "Jacob's Staff," or cross staff, a long shaft with a movable cross-

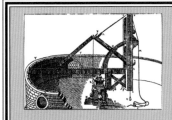

TRIQUETRUM

A triquetrum predates Ptolemy (100-170 A.D.), and was designed to measure the angular altitude of stars. It consists of three straight rods and resembles a triangle. The rods are linked, with a vertical bar in the center and a hinged rod at its top and bottom. The hinged rods can slide against each other, changing their point of intersection. One rod is marked with graduations, and the other is fitted with sights. To determine the altitude of a celestial body, the vertical bar is held perpendicular to the earth, while one rod is used to sight the object of interest. The angle between the graduated and sighted arms, or the position of these two arms against each other, provides the altitude.

Triquetrums ranged in size, and Copernicus was said to have built one that was eight-feet long.

ASTROLABE

The astrolabe was commonly made of brass and measured six to eight inches in diameter and a quarter inch in thickness. Its origins can be traced back to ancient Greece, and Arab astronomers refined it from the ninth through the eleventh centuries before introducing it to twelfth-century Europeans in Islamic Spain.

Affixed to the front of the astrolabe was an engraved disk with the North Pole as its center, and three rings representing the Tropic of Cancer, the equator, and the Tropic of Capricorn. (The southern hemisphere below the Tropic of Capricorn was not included.) Because it was latitude dependent, the plate included several disks that could be exchanged if the viewer changed latitudes. The circumference of the front was also inscribed with a time scale, often the twenty-four hours of the day.

A movable disk marked by significant heavenly bodies was placed on top of the engraved disk and represented the daily rotation of the sky. The astrolabe could then be used to determine the time of the day or night, predict the time of a celestial event (i.e. moonrise or sunset), and determine the position of objects in the sky at a particular time.

The back of the astrolabe was used for measuring the altitude of heavenly bodies. The astrolabe would be hung by the ring at its top, and the viewer would site the star or planet along a bar, or alidade, which was secured in the center of the disk and rotated. The observer could then read the altitude of the object from a scale engraved around the circumference of the disc.

The astrolabe also had two circular scales that were used to determine the Sun's position on the ecliptic (its yearly path in relation to the stars). One circle had the days of the year, and the other showed the corresponding locations of the Sun.

The astrolabe remained a popular astronomical instrument until the middle of the seventeenth century.

CROSS STAFF

A cross staff, also known as a "Jacob's Staff," is a simple tool that dates back to ancient times. Its cross shape is used to determine the angle between two objects. It consists of a long arm (staff) and a perpendicular shorter arm (cross piece) that can be slid up and down the staff. The viewer places one end of the long arm against the eye, with the staff pointing outward. The viewer moves the cross piece until the ends of the bar line up with the two objects in question, such as two celestial objects, or the horizon and a heavenly body.

The ratio of the length of the cross piece to the length measured on the staff provides the data necessary to determine the angle. If the longer arm were marked with a graduated scale, the viewer could determine the angle between the two objects by noting the point where the staff and the cross piece intersect. If the staff were marked with a linear scale, the ratios would be calculated using trigonometry.

bar that was used to calculate the angles of celestial objects relative to the horizon.

Although Copernicus made some of his own observations, he depended for the most part on the work of previous astronomers, including Ptolemy. He seems not to have seriously doubted the data's accuracy. In 1522, he was asked by Canon Bernhard Wapowsky of Cracow to review a new treatise, *On the Motion of the Eighth Sphere* by Johannes Werner. In his book, Werner challenged some of the observations made by Ptolemy and Timocharis, another ancient stargazer Copernicus depended on. Copernicus reacted with anger at the work. The observations of the ancients had been "bequeathed to us like an inheritance," he said, and he accused Werner of "slandering the ancients."

Following the circulation of *Little Commentary*, Copernicus continued to study and make mathematical calculations in an attempt to support his theories.

CHAPTER EIGHT
ON REVOLUTIONS

In the introduction to his masterwork, Copernicus explained why he had designed a new system of planetary motion. It troubled him that after centuries of work in astronomy there was still disagreement over what was correct. Most critically, none of the earlier models, including Ptolemy's, had successfully revealed planetary, or lunar, motion while maintaining uniform circular motion. Although he does not mention the equant specifically in the introduction, Copernicus had made it clear in the *Little Commentary* that he rejected its use.

Copernicus explained that his primary motivation was to restore the ancient ideal of unity. While pursuing this goal, he had become aware of earlier mathematicians who had suggested that the Sun was at the center of the planets and Earth was in rotation. He became willing to consider this alternative, absurd as it may have seemed, and had discovered that many of the problems were indeed solved by placing Earth in motion relative to the other planets. He did not give more specific information about what exactly had prompted him to make such a dramatic adjustment to his theories.

The manuscript, when finally published, was given the Latin title *De Revolutionibus Orbium Coelestium Libri VI*, or *Six Books On the Revolutions of the Heavenly Spheres*. Each of the six books addressed a specific aspect of Copernicus's theory. Preliminary materials included a preface, Cardinal Schoenberg's praise-filled letter urging Copernicus to publish his theory, and Copernicus's own letter, which dedicated his work to Pope Paul III.

Book One lays out Copernicus's basic argument: The stationary Sun is near the center of the planets, and the universe is bounded by the sphere of the fixed stars. Earth is in motion around the Sun, as are the other planets, in the following order: Mercury, Venus, Earth, Mars, Jupiter, and Saturn. Earth not only orbits the Sun, but also rotates on its own axis every twenty-four hours, which explains the apparent daily motion of the heavens, such as the rising and setting of the Sun. The Moon orbits Earth. Book One concludes with several chapters on plane and spherical geometry—mathematical tools Copernicus would use to support his arguments.

Book Two is a description of spherical astronomy and the armillary sphere, an ancient instrument comprised of several metal rings representing the celestial equator, the ecliptic (the apparent path of the Sun through our sky), and other circles of the heavenly sphere. Copernicus applies some of the geometrical rules laid out in Book One to spherical astronomy. The second book closes with a catalog of over a thousand stars and their positions defined by a pair of coordinates formed by the intersection of two circles on the celestial sphere. Copernicus borrowed this star catalog from Ptolemy, correcting a few errors but more or less reproducing it as it had been published centuries before in the *Almagest*. In fact, Copernicus used very few of his own observations in *On Revolutions*—just twenty-seven in a book of over four hundred pages.

Book Three is devoted to a discussion of the orbit of Earth and the length of the year. Book Four explains the motions of the Moon, its phases, and its eclipses. Books Five and Six address the movements of the planets. All of Books Two through Six are full of mathematical tables and geometrical sketches describing how the planets move around the Sun.

Copernicus faced the same problem that Ptolemy and others before him faced. Planetary orbits are not circular. However, In addition, much of the data he used was two thousand or more years old. He had to create devices to correct for these problems. He was able to do away

The title page of Copernicus's *On the Revolutions of the Heavenly Spheres.*

NICOLAI CO
PERNICI TORINENSIS
DE REVOLVTIONIBVS ORBI-
um coelestium, Libri VI.

Habes in hoc opere iam recens nato, & ædito,
studiose lector, Motus stellarum, tam fixarum,
quàm erraticarum, cum ex ueteribus, tum etiam
ex recentibus obseruationibus restitutos: & no-
uis insuper ac admirabilibus hypothesibus or-
natos. Habes etiam Tabulas expeditissimas, ex
quibus eosdem ad quoduis tempus quàm facilli
me calculare poteris. Igitur eme, lege, fruere.

ἀγεωμέτρητος οὐδεὶς εἰσίτω.

Collegij Brunsbergensis Societatis Jesu.

Norimbergæ apud Ioh. Petreium,
Anno M. D. XLIII.

with the equant only by replacing it with a smaller epicycle, or epicyclet, that was geometrically equivalent to the equant but moved in uniform circular motion.

Copernicus's system was not truly heliocentric because he claimed the actual center of the planetary orbits was not the Sun, but the center of Earth's orbit. The Sun, he said, did govern the planets, but he made little effort to explain the physical nature of the universe.

In many ways Copernicus was closer in spirit to Ptolemy than to the astronomers who came in his wake. His goal was to *revise* Ptolemy so that it was more in line with Aristotelian principles. Proposing that Earth moves was the radical step he took in pursuit of this goal. Copernicus has sometimes been referred to as the last medieval scientist and, in truth, modern astronomy was not born until seventy years later, when Johannes Kepler began analyzing planetary movements to develop the three laws of planetary motion—forever doing away with deferents, epicycles, equants, and epicyclets.

In chapter ten of Book One, Copernicus places the planets in their correct orbits and determines their orbital periods—the time it takes to complete one orbit. Using Earth's year as the baseline, the periods are as follows: Mercury, eighty-eight days; Venus, approximately seven and a half months; Mars, two years; Jupiter, twelve years; Saturn, thirty years. (At the time, these were the only known planets.)

He also correctly calculated the sideral year—orbital periods—of the known planets. He determined the pe-

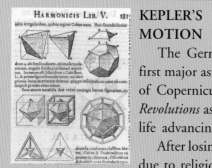

KEPLER'S LAWS OF PLANETARY MOTION

The German Johannes Kepler was the first major astronomer to accept the validity of Copernicus's system. After reading *On Revolutions* as a college student, he spent his life advancing the heliocentric system.

After losing his job teaching mathematics due to religious conflict between Catholics and Protestants, Kepler went to work for the Danish stargazer Tycho Brahe. After Tycho's death in 1601, Kepler inherited his vast store of astronomical observation. He used this data, the most accurate planetary observations ever recorded at that time, in his two major works. *The New Astronomy* and *World Harmony* attempted to find the mathematical harmony Kepler was convinced governed planetary motion. In the pursuit of this elusive goal, he discovered his famous three laws of planetary motion.

Although Kepler's three laws were founded upon two fundamental Copernican aspects—a central Sun and a moving Earth—they deviated from Copernicus by rejecting uniform circular motion. The more conservative Copernicus had been willing to rearrange the planets in order to maintain this Aristotelian precept. Kepler, however, used Tycho's observations to mathematically prove, in his first two laws, that orbits are elliptical, not circular, and that a planet's speed varies in inverse relation to its distance from the Sun. Kepler's third law established a correlation between a planet's orbital period and its distance from the Sun and would later prompt the Englishman Isaac Newton to develop his law of gravitation.

Kepler's Three Laws of Planetary Motion:
1. Planet's move in elliptical orbits, with the Sun at one focus.
2. A line drawn from the center of the sun to the center of a planet will sweep out equal areas in equal intervals of time.
3. The square of a planet's orbital period is proportional to the cube of its mean distance from the Sun.

ASTRONOMICAL UNITS

An astronomical unit (AU) is used by astronomers to measure the distances between the planets and the Sun. One AU is equal to 93 million miles, or the distance from Earth to the Sun. The chart below shows the distances of the other planets to the Sun.

Planet	Distance from the Sun	
	Astronomical Units	Miles (in millions)
Mercury	0.387	36
Venus	0.723	67.2
Earth	1	93
Mars	1.524	141.6
Jupiter	5.203	483.6
Saturn	9.539	886.7
Uranus	19.18	1,784.0
Neptune	30.06	2,794.4
Pluto	39.53	3,674.5

riod of Earth's sidereal year to be 365 days, 6 hours, 9 minutes, and 40 seconds. He was off by only thirty and one-half seconds. He might have still been thinking of his 1514 invitation to the Lateran Council when he stated this period was so constant it could be used for calendar reform. Copernicus also calculated the distance to the Moon as 60.30 of Earth's radii. (The radius of our planet is approximately four thousand miles.) Today we know the exact distance is 60.27 Earth's radii.

Copernicus provided a simpler solution to the cause of retrograde motion than the epicycle-deferent device. By placing Earth in orbit it was possible to understand the motion of the other planets relative to Earth. Because

Johannes Kepler analyzed the planets' movements to develop the three laws of planetary motion. *(Erich Lessing / Art Resource, NY.)*

Pope Gregory XIII

JULIAN AND GREGORIAN CALENDARS

The calendar used during Copernicus's life was commissioned by Julius Caesar in 48 B.C. Caesar had turned to Sosigenes, an astronomer from Alexandria, Egypt, to devise the new calendar. This Julian calendar consisted of 365 days grouped into twelve months, with the addition of an extra day every fourth year, or leap year. The lengths of the months alternated between thirty or thirty-one days, except February. To compensate for the existing discrepancies that had accumulated over the centuries, it was decided that the year 46 B.C. would include twenty-three extra days added to the end of February, and sixty-seven extra days between November and December, for a total of ninety extra days.

In 8 B.C., the Roman ruler Augustus changed the name of the summer month *Quintilis* to *Julius* (July) to honor Julius Caesar, and the summer month *Sextilis* to *Augustus* (August) to honor himself.

Originally, the Julian Calendar was not divided into weeks, but into days of business, *dies fasti*, and days when business was not conducted, *dies nefasti*. Weeks were not introduced until the fourth century A.D.

The Julian Calendar became the standard throughout the Roman Empire, and was adopted by the Christian church. Although more accurate than its predecessor, the length of the Julian Calendar did not coincide exactly with the true length of the year, which is 365.242199 days. Therefore, the Julian calendar gradually became out of synch with the seasons, just as the Roman calendar had, and the leaders of the Catholic Church knew action had to be taken. Several church councils were held, but no resolution to the problem was agreed upon. One major stumbling block was that astronomers did not know the exact length of the year and therefore had no confidence they would be able to definitively correct the current system.

In 1545, the Council of Trent attempted to take the situation in hand by asking Pope Paul II to devise a solution. Nothing was resolved until 1582, when Pope Gregory XIII issued a papal bull (decree) instituting changes. The new calendar, known as the Gregorian Calendar, was given 365.2422 days, and in order to right the dates, ten days were taken out of October of that year. In the Gregorian system leap years still occur every four years except in century years that are not also divisible by four hundred.

Soon after Pope Gregory's papal bull, most Catholic regions adopted the new calendar. It took more than a hundred years before most Protestant regions accepted it. Britain and its colonies came on board in 1752. When a region accepted the calendar, it lost approximately ten days. Some countries waited much longer to adopt the Gregorian Calendar, including Japan, which did so in 1853, and Turkey, which made the change in 1927.

The change from the Julian calendar to the Gregorian calendar is significant to the study of history because dates that fall before October 15, 1582, and after 46 B.C., have been recorded in the Julian system.

Earth is in third position, Mercury and Venus "lap" it on a regular basis, and Earth "laps" the outer planets. This was most noticeable with Mars, which appears to stop its progression and back up when it is in opposition. In effect, Earth is passing it on the inside lane of the orbital track, and this is what gives the impression of a stop-back-stop-forward motion as viewed from Earth.

Copernicus also made some remarkably accurate mathematical calculations. In his attempt to compute the dimensions of our solar system, he used what is known today as an Astronomical Unit, or AU. One AU equals the distance from Earth to the Sun. Copernicus determined Mercury is 0.38 AUs from the Sun, Venus 0.72, Mars 1.52, Jupiter 5.22, and Saturn 9.17. These values are very close to the actual values.

He did get some things wrong, however. His orbital sizes were correct when expressed in AUs but when he tried to calculate distances using Earth's radii, he missed the mark. Because he miscalculated the radius of Earth's orbit as 1142 of Earth's radii—over twenty times too small—a ripple effect ran through all of his calculations of planetary distances. Saturn's orbit, as figured by Copernicus, would be somewhere inside the orbit of Mercury. When he tried to calculate the diameter of the Sun, he arrived at a figure only five and a half times the diameter of Earth. The Sun's diameter is actually more than one hundred times that of Earth's.

One of the problems with his measurements was his incorrect calculation of the parallax of the Sun. As explained earlier, parallax is the phenomenon of apparent change in the relative position of an object due to a change in the observer's position. For example, hold a pencil out in front of you at arm's length and close one eye. Next, open the closed eye and close the open eye. Not only will the pencil appear to shift back and forth, the background will as well. Calculate the angle of change and you have determined the pencil's parallax. This method of calculation is known as triangulation, and is used by engineers when building roads or bridges and by surveyors, who measure plots of land by calculating distances to prominent landmarks.

Triangulation is easy when calculating distances on the surface of Earth. You make two measurements in order to form a baseline, which gives you the value for one leg of a triangle. Then you determine the values from each end of the baseline to the third object and calculate all of the distances involved with trigonometry. Even with modern instruments, it is a much more difficult task to make an accurate baseline measurement to a distant object in space from a rotating planet that is traveling in a huge ellipse. Imagine what a task it was for a sixteenth century scientist, with only the crude instruments Copernicus possessed.

The question of parallax was central to arguments for and against the movement of Earth, as it had been in ancient times. If Earth is moving, said the critics of heliocentricism, why had no one been able to measure stellar parallax—the apparent shift of the stars' locations? In chapter six of Book One, Copernicus answers, "The heavens are immense in comparison with the Earth and present the aspect of an infinite magnitude, and that in the judgement of sense-perception the Earth is to the heavens as a point to a body and as a finite to an infinite magnitude." Simply stated, the stars are so far away that no parallax is noticeable. Earth's orbit is so small relative to its distance from the stars that Earth, even in its motion, is just a tiny pinprick when compared to the vastness of space. This was one of the most revolution-ary ideas presented in *On Revolutions*. Copernicus believed the universe to be immensely larger than ever thought possible. He stopped short of saying there was no outer stellar sphere—that space was infinite—but he did argue that the stars were so far away that it was practically indistinguishable from infinity.

Stellar parallax was not measured until the nineteenth century. Even today, such measurements are difficult to make and can only be calculated for stars within several hundred light years of us because, even if the entire orbit of Earth is used as a baseline, that length is minute compared to the distance of even our nearest star.

Copernicus also used the enormous size of the universe to refute arguments against Earth's two motions. One argument was that if Earth were spinning everything would fly off of it due to centrifugal force, the way a ball will fly off a merry-go-round. Because we do not all fly into space, Earth must be stationary. Copernicus cleverly turned this argument back on his critics. If Earth is not rotating, he said, then the entirety of the heavens must be spinning around at near-infinite speed, given the distance they are from Earth and the fact that they make an entire revolution every twenty-four hours. Which is more plausible, he asked, Earth spinning around at some yet to be determined velocity, or the entire universe spinning at a huge multiple of Earth's velocity? To explain why people and objects did not fly off into space, Copernicus proposed that the layer of air surrounding Earth is carried along as it turns. Therefore, objects in the atmosphere would also be carried along, as were birds and clouds.

On Revolutions is an amalgam of brilliant mathematical work and insight wedded to ancient ideas that could no longer be supported by facts. His adherence to the ideal of uniform circular motion distorted his heliocentric model just as it had Ptolemy's geocentric model. While he correctly calculated many values about the size of our solar system, he was wrong on others. Although it had limitations, Copernicus had written a remarkable book that would lead to the creation of modern astronomy. In the beginning, the influence of his work was limited, creating only the smallest of pressures against the wall of medieval philosophy. Over time, the force of his basic propositions—that Earth and the other planets orbited a stationary Sun while also spinning on an axis—would replace not only what we know about the solar system, but how we understand our place in the universe, as well.

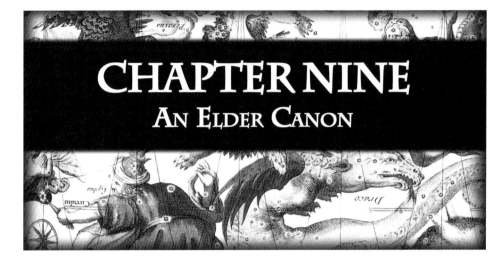

CHAPTER NINE
AN ELDER CANON

Scholars have speculated for centuries about what made Copernicus hesitate so long before publishing his work. He mentioned one concern in the dedication of *On Revolutions* addressed to Pope Paul III:

> I can well appreciate, Holy Father, that as soon as certain people realize that in these books which I have written about the Revolutions of the spheres of the universe I attribute certain motions to the globe of the Earth, they will at once clamour for me to be hooted off the stage with such an opinion. For I am not so pleased with my work that I take no account of other people's judgement of it.

Although he was a modest man, Copernicus lived, to a certain degree, a public life as a canon. Publishing his work could lead to deeper involvement in public life. While it is speculative to say, it is possible he was torn between pride in his life's work and the desire to live out his life as undisturbed as possible.

In sixteenth century Poland, there was little distinction between politics and religion. The Protestant Reformation had begun in 1517, a few hundred miles to the south. By the time Copernicus had completed the majority of his writing, the schism in European Christianity had already spread throughout most of what is today northern Germany and into Scandinavia. Although Grand Master Albert, the leader of the Teutonic Knights, had converted to Protestantism after becoming the duke of Prussia in 1525, the king of Poland had remained faithful to the Roman church and had sent troops to suppress a Lutheran rebellion in Gdansk.

Following the death of Copernicus's Uncle Lucas, Fabian von Lossainen was made bishop of Varmia. He was a tolerant man who had called Luther "a learned monk who has his own opinions regarding the scriptures." Unfortunately for Copernicus, the next bishop was a much different leader. One of Mauritius Ferber's first actions upon being named bishop of Varmia was to issue an edict stating that anyone who converted to Lutheranism would be "cursed for all eternity and smitten with the sword of anathema." This pronouncement created conflict within Poland. In reaction, the bishop

Copernicus dedicated *On Revolutions* to Pope Paul III, pictured here in a portrait by Titian from 1543. *(Courtesy of Museo Nazionale di Capodimonte, Naples, Italy.)*

of Samland, a neighboring diocese, published an edict recommending his clergy read Luther's writings. Not only that, he gave permission for them to preach and baptize in the common languages (instead of Latin), a central tenet of Lutheranism. The beginning of a centuries-long conflict was brewing, a potent draught that would result in warfare, great suffering, and destruction.

Copernicus's closest friend, Canon Tiedemann Giese, published a tract pleading for tolerance. He ends his work with this statement: "Oh, if only the Christian spirit informed the Lutheran attitude to the Romans, and the Romans' toward the Lutherans—verily, then our

Churches would be spared these tragedies of which no end can be seen . . . Verily, the wild beasts deal more kindly with each other than Christian deals with Christian." Giese mentions Copernicus in a prefatory letter to the tract, and Copernicus must have given his friend permission to mention him in the text, which would suggest to the reader that Copernicus approved of the book and its message of tolerance. There can be little doubt that many more of the older humanist clergy felt the same way about the issue.

Around the time Copernicus finished his book, a new bishop was installed at the diocese of Chelmno. Johannes Dantiscus, whose original name was Johannes Flachsbinder, was the son of a brewer in Gdansk, the city from which he took his Latin name. Dantiscus had led an exciting life. He fought in the Polish military against the Tartars and the Turks. He attended the University of Cracow and traveled extensively in Greece, Italy, Arabia, and the Holy Land. By the time he was twenty-three, he was working as a special envoy for the king of Poland, after having served as his secretary. During this period, he met Copernicus, who was then working with Uncle Lucas. While Copernicus was settling down in Varmia and turning his attention to astronomy, Dantiscus was promoted to Polish ambassador to the Holy Roman Emperor Maximilian, as well as to his successor, Emperor Charles V. Maximilian respected him so much that he appointed him poet laureate and knighted him; Charles gave him a Spanish title. Both borrowed him from the king of Poland and sent him on diplomatic missions to Venice and Paris.

At the age of forty-five, Dantiscus decided to retire from politics and diplomacy and arranged to be appointed bishop of Chelmno in 1532. He was also made a canon of Frombork Cathedral, which made him a *confrater*, or colleague, to Copernicus. Bishop Dantiscus corresponded with many scholars throughout Europe and tried to strike up a friendship with Copernicus. He wrote to Copernicus several times and was apparently interested in his astronomical research. For some unknown reason, Copernicus did not respond to Dantiscus's overtures of friendship.

Dantiscus was elected bishop of Varmia on September 20, 1537, following the death of Mauritius Ferber. His election was a foregone conclusion, given his close association with the king of Poland. There was one small bit of intrigue associated with Dantiscus's election that also involved Copernicus. Originally, the short list of candidates for the position included Canons Zimmerman and Heinrich Snellenberg. However, Snellenberg had borrowed a hundred marks from Copernicus and, after twenty years, had only repaid ninety of it. Because of this, Copernicus's friend Tiedemann Giese wrote Dantiscus a letter requesting that Snellenberg be taken off of the list and replaced with Copernicus. Dantiscus had no strong feelings on the matter and obliged Giese's request. Copernicus could then say he had once been considered as a candidate for bishop of Varmia.

The cordial relations between Copernicus and Dantiscus did not last, though. Shortly after becoming bishop in the autumn of 1538, Dantiscus made a tour of all of Varmia's towns in the company of two of his canons, one of whom was Copernicus. During this trip, he broached a delicate subject concerning a woman named Anna Schilling, a distant cousin of Copernicus. Schilling was Copernicus's *focaria*, a word that had a dual meaning—either housekeeper, or concubine, or both. Apparently, Copernicus and one of the other canons, Alexander Sculteti, were both openly living with their women housekeepers. Sculteti had several children by his *focaria*.

Dantiscus was a bishop in the Catholic Church at a time when much was changing. In response to the spread of Protestantism, the Church was in the early stages of what came to be called the Counter-Reformation. The Church was trying to reform itself in many areas while also developing new orders, such as the Society of Jesus, or Jesuits, who would help to spread the Catholic message. A canon living in a semi-open sexual relationship with his *focaria* had not been uncommon in years past. Now, however, as the church was entering into a new era and had to compete with the growing sect of Protestantism, it was no longer so easy to ignore this practice. After returning from the trip, Dantiscus ordered Copernicus and Sculteti to send the women away.

Copernicus was now sixty-three and reluctant to change. He wrote to Dantiscus, however, and promised to send Anna away soon, but she refused to leave. Sculteti's housekeeper reacted in the same way.

Shortly after this, a Canon Plotowski wrote the bishop with a report on the "Frombork wenches." He claimed that Sculteti's *focaria* had "hid for a few days in his house." Plotowski described her as "a beer-waitress tainted with every evil." In the same letter, he reported, "the woman of Dr. Nicholas did send her things ahead to Gdansk, but she herself stays on in Frombork."

Six months later, the matter remained unresolved. Dantiscus grew tired of sending orders to Copernicus only to receive letters filled with praise and false promises in reply. He asked Copernicus's friend Giese, who was now bishop of Chelmno, to intervene. Giese was firmly on Copernicus's side. He wrote back that the reports that Anna and Copernicus were secretly meeting were "trumped up charges."

Certainly Copernicus saw the Catholic Church was becoming more restrictive and the tolerant attitude that had existed during his youth was ending. The fight against Protestantism was injecting a higher level of militancy both within the Church and in its dealings with the outside world. Church corruption was one of the driving forces of the Protestant Reformation. Copernicus was caught, fairly or unfairly, in the effort to curtail it.

It was at this point, when the conflict between Dantiscus and Copernicus was approaching the critical stage, that a new development arose in the generally quiet life of Nicholas Copernicus. A brilliant young mathematician named Georg Joachim von Lauchen came to Frombork.

Georg Joachim von Lauchen had taken the Latin name Rheticus after his father was convicted of sorcery and beheaded when Georg was fourteen. A brilliant student, Rheticus had studied in Italy and at German universities. In 1536, at age twenty-two, he was appointed lecturer in mathematics at the University of Wittenberg. Three years later, he requested a leave to travel to Frombork. Rheticus probably learned about Copernicus from his teacher Johann Schoener in Nuremberg. He became determined to meet the scientist and discuss his writings with him.

Rheticus, a Lutheran, was willing to entertain the idea of heliocentricism, despite the fact that Martin Luther, who was still alive, had rejected the theory. Luther's powerful friend and associate Philipp Melanchthon, who was primarily responsible for establishing a Protestant educational system, was a bit more tolerant of the idea. Melanchthon's attitude was the main reason Rheticus was allowed to visit Copernicus.

Before Rheticus's arrival in Frombork, Bishop Dantiscus published an edict forbidding Lutherans from traveling in Varmia. This, however, did not stop the determined scholar from making the trip. Soon after Rheticus's arrival, he and Copernicus traveled to Chelmno to visit with Bishop Giese. The decree was most likely the impetus for this trip, since Dantiscus's edict was limited to his own diocese of Varmia.

CHAPTER TEN
PUBLICATION

Rheticus and Copernicus spent most of the summer and fall of 1539 with the bishop of Chelmno, Tiedemann Giese, at Castle Loebau, his official residence. Giese and Rheticus soon formed an unlikely duo, a Lutheran intellectual and a Catholic bishop, working together in an attempt to convince Copernicus to publish his life's work. Copernicus frustrated them by making a series of excuses as to why it should remain unpublished. At one point, he agreed to publish a new set of astronomical tables based on his model but would not include any discussion of his heliocentric theory. That way, only those learned in mathematics and astronomy would be able to discern the importance of what he was proposing. Giese opposed this idea, stating, "There was no place in science for the practice frequently adopted in kingdoms, conferences, and public affairs, where for a time plans are kept secret until the subjects see the fruitful results." Giese diverted Copernicus away from his fear of public ridicule by saying the average uneducated citizen would never understand his ideas, much less be able to decipher the mathematical explanations.

A compromise was eventually reached. Rheticus would write a paraphrased version of the manuscript explaining its contents. In the manuscript, Rheticus referred to Copernicus as *domine praeceptor* (master teacher) and as "the learned Dr. Nicholas of Torun."

Rheticus and Copernicus returned to Frombork, where the young scholar spent ten weeks analyzing and writing a treatise on his teacher's masterwork. Rheticus's book was published as *Narratio Prima de Libris Revolutionum* (*First Account of the Book of Revolutions*). It was written in the form of a letter from Rheticus addressed to Johannes Schöner, who was probably the man who first introduced him to the ideas of Copernicus. This had been a common literary device since ancient times; many books of the New Testament, for example, are letters addressed to an individual but composed for a general audience. Rheticus stated in *First Account* that he was writing this first account of the material and would later write a second account that would explain the portions he did not yet fully understand.

Rheticus began his summary by pointing out that the work was composed of six books. From this point, though, he did not follow the structure of Copernicus's

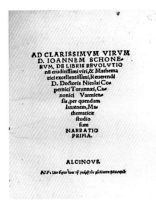

AD CLARISSIMVM VIRVM
D. IOANNEM SCHONE-
RVM, DE LIBRIS REVOLVTIO
nŭ eruditiſſimi viri,& Mathema
tici excellentiſſimi,Reuerendi
D. Doctoris Nicolai Co-
pernici Torunnæi,Ca-
nonici Varmien-
ſis ,per quendam
Iuuenem,Ma-
thematicæ
ſtudio
ſum
NARRATIO
PRIMA.

ALCINOVS.

Rheticus published the *Narratio Prima*, or *First Account*, of Copernicus's theory.

book. Instead, he arranged the work according to what he saw as the essential ideas. He also included a lengthy astrological section in which he claimed changes in Earth's orbit were responsible for the rise and fall of both the Roman and the Muslim empires, and would usher in the second coming of Christ. He also estimated that the world would only survive for six thousand years, and he included quotations from Aristotle and Plato.

Rheticus included this additional material for two reasons. First, he believed in astrology, as did many of the people who would read his work. (We do not know how much faith Copernicus put in astrology, but the fact that *On Revolutions* does not contain astrological material among its theories of the heavens most likely indicates that he was not a deep believer.) Second, Rheticus knew, as did Copernicus, that many scholars would reject his ideas outright because they were in conflict with Aristotle and Ptolemy. Rheticus made comments throughout the text that were intended to soften the blow Copernicus's theories were delivering against the ancient understanding of the universe.

First Account was completed by the end of September 1539. At that time, Rheticus left Frombork and set out for Gdansk, the nearest town where he could have the work printed. The first copies were shipped out of Gdansk in February 1540. Rheticus sent a copy to Philipp Melanchthon, Luther's closest friend and assistant, in Wittenberg. Another of Rheticus's scholarly friends, Achilles Permin Gassar, arranged for an edition to be printed in Basel, Switzerland, only a few weeks after the Gdansk edition was issued. Giese sent a copy of *First Account* to Duke Albert of Prussia, who, although a Lutheran, was an early supporter of Copernicus's ideas. This double publication helped to speed the dissemination of *First Account*. Within months, Rheticus and Giese were not the only voices encouraging Copernicus to publish his entire work.

Meanwhile, Rheticus had returned to Wittenberg to resume his teaching duties. At the end of the summer, he journeyed back to Frombork with the intention of writing the "Second Account" and to continue trying to persuade Copernicus to publish the entire book. Copernicus finally agreed to release *On Revolutions*, provided Rheticus oversaw its publication.

Rheticus stayed at Frombork over a year, from the summer of 1540 until September 1541. He spent the majority of his time copying the manuscript by hand and verifying most of the mathematical calculations. He made some corrections and minor alterations. In addi-

PHILIPP MELANCHTHON

A friend and fellow reformer of Martin Luther, Philipp Melanchthon worked to systematize the theology of the Reformation, establish Protestant public education, and define the Lutheran statement of faith. Born in 1497, in what is now Switzerland, he worked for forty-two years at the University of Wittenberg. He studied theology, but was never ordained and did not preach. Melanchthon wrote the first treatise on "evangelical" doctrine. The treatise deals primarily with practical religious questions: sin and grace, law and gospel, justification and regeneration. This popular work ran through more than one hundred editions before his death.

In 1518, at the age of twenty-one, he became the first professor of Greek at the University of Wittenberg. He was a favorite of the students and established a deep friendship with Luther. Ten months before Melanchthon began teaching, Luther had nailed his ninety-five theses to the church door. This act established the arc of his life as well as Luther's. Melanchthon spent the rest of his life helping to define, systematize, and defend Protestantism.

In 1530, Melanchthon wrote a statement of faith that was presented to the Diet of Augsburg. Known as the Augsburg Confession, he attempted to prove, despite differences of opinion, that the beliefs of Luther and his followers were within the general teachings of the Catholic Church. The confession was rejected by the Church of Rome, but became the creed of the Lutheran Church.

Luther thought so highly of Melanchthon's contribution to the Protestant movement that he wanted him to become its leader after his death. Melanchthon's leadership was controversial, however, because of infighting within the movement. Many believed Melanchthon did not go far enough in his theology and worked too hard to find common ground with Catholics. His place in Protestantism was reviled until the eighteenth century.

tion, he did several other tasks for the aging canon. Copernicus had let lapse a project to create a map of Varmia. Rheticus finished the task for him, not only completing the map, but also composing a gazetteer and a treatise on cartography to go with it. He sent it all to Duke Albert of Prussia as gifts, with a letter of dedication giving prominent mention to Copernicus's forthcoming book.

Giving the map and other gifts to Duke Albert were not merely acts of generosity. Rheticus wanted to please the duke because he wished to enlist his help in the printing of *On Revolutions*. He asked Duke Albert to write two letters recommending publication. One letter was addressed to the Protestant elector of Saxony and the other to the University of Wittenberg. Rheticus wanted to have the books printed at a very well known shop that specialized in works of astronomy, located in the Protestant city of Nuremberg. Rheticus was worried that because Martin Luther was opposed to the heliocentric theory, he might have problems getting the work published in a Protestant city. Fortunately, though, the letters from Duke Albert were enough to convince the authorities that it would be fine to print the text.

Rheticus now had everything he needed to finally publish Copernicus's life's work, except time. He finished copying the original manuscript in August 1541, and returned to Wittenberg for a visit. However, upon his return, he was elected dean of the faculty. It was a promotion he could not turn down without seriously

Nicholas Copernicus introduced one of the ideas that helped the development of modern science.

damaging his career. Rheticus was forced to postpone his editorial and publishing duties until the end of the academic year in May 1542, at which time he finally left for Nuremberg.

As the preliminary arrangements for publication were begun, Rheticus received word of yet another career advancement, this time to the chair of mathematics at the University of Leipzig. He had applied for the prestigious position earlier in the year and had to leave for Leipzig before the printing of *On Revolutions* was finished, in order to take up his new duties. After all of the time and effort he had put into bringing *On Revolutions* to publication, he had to leave before the task was completed.

Before he left for Leipzig, Rheticus turned the responsibility for overseeing the publication of *On Revolutions* to a trusted friend and noted theologian. Andreas Osiander had been one of the earliest converts to Lutheranism but was also a supporter of Copernicus's ideas. He had even corresponded with him in years past. However, Osiander was worried over how the book would be received. In order to deflect criticism, he decided, without informing Rheticus or Copernicus, to write a preface that stated, "These hypotheses need not be true or even probable" but merely sufficient to predict the movements of the planets. That they can appear to be true is enough, he said. He concluded the preface by saying that these new concepts deserved to be known "together with the ancient hypotheses which are no more probable."

The most damaging thing about the preface is that Osiander did not sign it, which gave the impression that it had been written by Copernicus. Although Copernicus's own introduction only two pages later contradicts Osiander's statement, most readers concluded Copernicus himself had not been convinced of the truth of his argument and was only presenting it as a hypothesis. It would be decades before the true author of the introduction was revealed.

In Osiander's favor, he undoubtedly included his preface with good intentions. More than two years earlier, in the period before Copernicus had decided to publish his work, Osiander had written a letter, which he sent to both Copernicus and Rheticus, stating:

> For my part I have always felt about hypotheses that they are not articles of faith but bases of computation, so that even if they are false, it does not matter, provided that they exactly represent the phenomena . . . It would therefore be a good thing if you could say something on this subject in your preface, for you would thus placate the Aristotelians and the theologians whose contradictions you fear.

However, Osiander's good intentions led, fairly or not, to his having an infamous reputation in science history. Tiedemann Giese was especially furious when he read the preface. He wrote Rheticus a letter urging him to appeal to the Council of Nuremberg to intervene in the matter. Rheticus, for unknown reasons, did little about the situation.

Whether Copernicus knew of the preface is unknown, although there has been speculation that he did know of it. Giese indicated his friend had received galleys of the book from both Rheticus and Osiander as it was being prepared for press. However, Copernicus was by this time a very sick man. We do not know for certain that there was a rush to get a copy of the finished book to Copernicus before he died. We do know that he was entering the last weeks of his life as his book was being published in far away Nuremberg. He had suffered a cerebral hemorrhage in November 1542, and the last six months of his life were spent in bed. Giese recorded his final moments in a letter to Rheticus: "For many days he had been deprived of his memory and mental vigour; he only saw his completed book at the last moment, on the day he died." Nicholas Copernicus died on May 24, 1543. He was seventy years old. We do not know for certain that he ever saw a copy of the book that would begin what has been called the Copernican Revolution.

CHAPTER ELEVEN
THE RETREAT OF THE STARS

It took many years before the basic truth of Copernicus's theory was accepted by most of Europe's astronomers. While only a small number of *On Revolutions* was printed in 1543, it was reprinted several times over the next centuries. Until the nineteenth century, the reprints were taken from the 1543 edition, and it was not until 1854 that Osiander's inserted preface was removed from the text.

The fact that so few copies had been printed in the first edition was only one of the reasons that Copernicus's heliocentric theory was slow to be accepted. The book itself was very difficult to read, as most of it was filled with astronomical mathematics.

Some of the astronomers and mathematicians who

could read the book, and did, were profoundly changed by it, even when they did not agree with its basic tenets. Erasmus Reinhold, a colleague of Rheticus and professor of mathematics at the University of Wittenberg, may not have completely accepted Copernicus's theory. He did, however, use the computations in *On Revolutions* to produce a set of planetary tables based on the heliocentric model that he published in 1551. Duke Albert of Prussia paid for the printing of the tables, known as the *Prussian Tables*. The *Prussian Tables* were generally more accurate in their predictions than any previously produced and were used until Johannes Kepler's *Rudolphine Tables* of 1627 superseded them.

It is not surprising that the theory of a moving Earth and central Sun was slow to be accepted. It was contrary to common sense and seemed to be physically impossible. Copernicus had attempted to explain how Earth could be spinning in two motions at the same time without slinging all of its inhabitants, not to mention buildings and trees, off into space. But he was not a physicist. It would not be until Galileo Galilei used the experimental method to rewrite the laws of physics that an orbiting Earth began to seem possible.

As time passed, though, nature itself began to make the case for Copernicanism. Aristotle had said the heavens were perfect and unchanging, but during the decades after Copernicus's death, several cosmic events occurred that contradicted Aristotle. Later in the sixteenth century, the Danish astronomer Tycho Brahe

TABULÆ

RUDOLPHINÆ,

QUIBUS ASTRONOMICÆ SCIENTIÆ, TEMPO-
rum longinquitate collapsæ Restauratio continetur;

A Phœnice illo Aſtronomorum

TYCHONE

Ex Illuſtri & Generoſa BRAHEORUM in Regno DANIÆ
familiâ oriundo Equite,

PRIMUM ANIMO CONCEPTA ET DESTINATA ANNO
CHRISTI MDLXIV: EXINDE OBSERVATIONIBUS SIDERUM ACCURA-
TISSIMIS, POST ANNUM PRÆCIPUE MDLXXII, QUO SIDUS IN CASSIOPEIÆ
CONSTELLATIONE NOVUM EFFULSIT, SERIO AFFECTATA: VARIISQUE OPERIBUS, CUM ME-
chanicis, tum librariis, impenſo patrimonio ampliſſimo, accedentibus etiam ſubſidiis FRIDERICI II. DANIÆ
REGIS, regali magnificentia dignis, tracta per annos XXV, potiſſimum in Inſula freti SUNDICI HUEN-
NA, & arce URANIBURGO, in hos uſus à fundamentis extructâ :

TANDEM TRADUCTA IN GERMANIAM, INQUE AULAM ET
Nomen RUDOLPHI IMP. anno M D IIC.

TABULAS IPSAS, JAM ET NUNCUPATAS, ET AFFECTAS, SED
MORTE AUTHORIS SUI ANNO MDCI DESERTAS,

JUSSU ET STIPENDIIS FRETUS TRIUM IMPPP.
RUDOLPHI, MATTHIAE, FERDINANDI,
ANNITENTIBUS HÆREDIBUS BRAHEANIS; EX FUNDAMENTIS
Obſervationum relictarum; ad exemplum ferè partium iam exſtructarum; continuâ multorum annorum
ſpeculationibus & computationibus, primùm PRAGÆ Bohemorum continuavit; deindè LINCII,
ſuperioris Auſtriæ Metropoli, ſubſidiis etiam Ill. Provincialium adjutus, perfecit, abſolvit,
adq; auſarum & calculi perennis formulam traduxit

IOANNES KEPLERUS,

TYCHONI *primùm à* RUDOLPHO II. *Imp. adjunctus calculi miniſter; indeq;*
trium ordine Imppp. Mathematicus :

Qui idem de ſpeciali mandato FERDINANDI II. IMP.
petentibus inſtantibusq; Hæredibus,

Opus hoc ad uſus præſentium & poſteritatis, typis, numericis proprijs, cæteris
& prælo JONÆ SAURII, *Reip. Ulmanæ Typographi, in publicum*
extulit, & Typographicis operis ULMÆ *curator affuit.*

Cum Privilegiis IMP. & Regum Rerumq; publ. vivo TYCHONI ejusq; Hæredibus,
& ſpeciali Imperatorio, ipſi KEPLERO conceſſo, ad anno XXX.

ANNO M DC XXVII.

Johannes Kepler completed the Rudolphine Tables after Tycho Brahe died.

Although he made many observations of the planets and stars, Danish astronomer Tycho Brahe rejected the heliocentric model and proposed a geocentric one of his own design. *(Courtesy of The Royal Library, Copenhagen.)*

attempted to calculate the parallax of a comet and discovered he could not. This could only mean that the comet traveled beyond the Moon, in the supposedly immutable celestial sphere, an undeniable contradiction of Aristotle.

Tycho Brahe, along with the rest of the world, witnessed another change in the sky that cut the ground from beneath Aristotle. In November 1572, a supernova suddenly appeared in the constellation of Cassiopeia. A supernova is a star that has exploded and expels vast amounts of energy in the form of hot gases and light. The "new star" of 1572 was so bright that it was visible during the day. It remained observable to the naked eye—telescopes had not yet been invented—until the spring of 1574. Another supernova appeared in 1604, and was observed by Johannes Kepler. Both of these supernovae were classified as Type I, the explosion of a white dwarf star.

Although this type of supernova happens frequently, it is very rare for them to occur that close to Earth. Hipparchus, an ancient Greek astronomer, claimed to have witnessed a "new star" in the second century B.C. that was actually a supernova, and there are reports that one was visible from China in 1006. The two supernovae, sighted only thirty years apart, are the most recent to have been observed in our Milky Way galaxy. It is

Galileo used the recently invented telescope to discover Venus's four phases as it orbits the Sun, providing the first physical confirmation of Copernican theory. *(Courtesy of the National Maritime Museum, London.)*

remarkable that these rare celestial events occurred so close together just as Copernicus's ideas were being widely debated.

In the face of this mounting astronomical evidence, philosophers and scientists began to consider Copernicanism with a more open mind. In 1573, thirty years after Copernicus's death, the English mathematician Thomas Digges published a small book about the 1572 supernova. Digges had tried to calculate its parallax and failed, as had Tycho Brahe. But unlike Brahe, who never accepted the idea of a moving Earth, Digges devoted a section of his book to attack the Ptolemaic system as

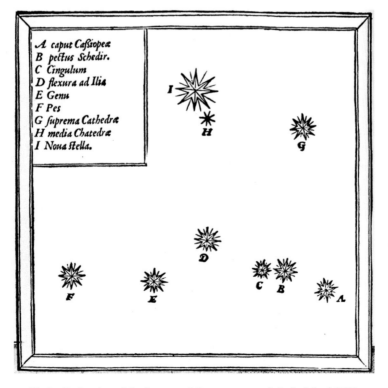

Tycho Brahe drew this diagram of the supernova (labelled I) of 1572.
(Courtesy of The Royal Library, Copenhagen.)

Johannes Kepler worked with Tycho Brahe in the years preceding the Dane's death. Kepler inherited his predecessor's data and used it to found the science of physical astronomy. *(Courtesy of the Sternwarte Kremsmünster, Austria.)*

being overly complex, and to defend Copernicus's approach as more logical. In a 1576 work, Digges reproduced Copernicus's arguments for a moving Earth and a stationary sky. He modified the theory by pushing the stars out to greater distances and proposed they were scattered at random in an infinite universe no longer bound by any "crystal spheres." The 1576 work contains a diagram of the celestial system, with the Sun at the center and the planets ordered as Copernicus had laid out. However, instead of an eighth sphere containing the stars and beyond which was nothing, Digges places his stars around the drawing all the way to the edge of the page as a visual representation of the last sphere, which is an "orbe of starres fixed infinitely up extendeth hit self in altitude spherically." The stars, Digges concludes, go on forever. The controversial idea of an infinite universe was beginning to attract supporters only thirty years after the publication of *On Revolutions*.

In 1600, the Italian monk Giordano Bruno was burned at the stake, partly because of his public profession of belief in the Copernican system. Bruno believed that the Sun was only one star in an infinite number of stars, each with its own planetary system of satellites. He wrote that

Giordano Bruno was burnt as a heretic in 1600 partly for his support of the Copernican system. *(Courtesy of the British Museum.)*

this pattern was repeated throughout endless space. Bruno's argument for an infinite universe was theological, not astronomical. He believed that because God possessed unlimited power, he would express himself by creating an infinite world.

Copernicus lived in the gap between ages. On one hand, he is an example of the Renaissance humanist who embraced knowledge for its own sake and was willing to reconsider old beliefs. When he realized that postulating the Sun as the focal point in the system of planets made it easier to account for retrograde motion and to solve other seemingly unexplainable astronomical problems, he was willing to consider the possibility with an open mind. On the other hand, his inability to see beyond Aristotle's philosophy, especially the principle of uniform circular orbits, reflects the older medieval mindset that valued loyalty to handed-down ideas over empirical evidence.

Although he may not have been able to step unhindered into the modern era, Copernicus possessed both insightful intelligence and moral courage. He propelled the human race on its long journey from the false security of a stationary, central Earth to acceptance of Earth's more humble role as one of the planets orbiting a single star in the ceaseless motion of the heavens.

TIMELINE

1473—Nicholas Copernicus is born in Torun, Poland, on February 19.

1483—Nicholas and his siblings are adopted by their Uncle Lucas Watzenrode following the death of their father.

1489—Uncle Lucas is appointed bishop of Varmia.

1491—Nicholas and his brother, Andreas, begin studies at the University of Crakow. Nicholas adopts the Latinized name Copernicus.

1494—Copernicus is called away from his studies in the hopes that his Uncle Lucas can appoint him to a position as a canon at Frombork Cathedral. This, however, does not occur.

1496—Copernicus leaves Varmia to study canon law in Bologna, Italy.

1497—Uncle Lucas has Copernicus appointed to a canonry at Frombork. Copernicus and his teacher, Domenico Maria Da Novarra, observe the moon approach Aldebaran, a double-star in the constellation Taurus.

1500—Copernicus and Andreas attend the Jubilee in Rome. Copernicus observes an eclipse of the moon.

1501—Copernicus and his brother return to Frombork to be officially installed as canons on July 27. Copernicus returns to Italy to study medicine at the University of Padua.

1503—Copernicus receives his Doctorate of Law at the University of Ferrara and returns to Varmia to begin working for his Uncle Lucas.

1509—Copernicus circulates the *Commentariolus*.

1512—Uncle Lucas dies. Copernicus takes up his position as canon at Frombork Cathedral.

1514—Copernicus is invited to send his opinion to the Lateran Council on calendar reform but declines to do so.

1516—Copernicus is sent to take up duties as the new administrator for Olsztyn and Pieniezno, two districts in the south of Varmia some fifty miles from Frombork.

1517—Martin Luther starts the Protestant Reformation.

1519—War begins between the Polish king and the Teutonic Knights.

1520—Frombork is besieged by the knights early in the year. In the fall, Copernicus returns to Olsztyn and defends it against a siege by the knights.

1521—An armistice is signed and the war ends. Copernicus is made commissar of Varmia, a position which gives him the authority to oversee the rebuilding following the war.

1530—Copernicus probably finishes the main portions of *On Revolutions*.

1539—Rheticus arrives in Frombork to see Copernicus about his theories and to convince him to publish his book.

1540—*Narratio Prima* is published in Gdansk. Copernicus finally agrees to publish his book.

1543—*On Revolutions* is published. Copernicus dies on May 24.

SOURCES

Chapter One: Born at the Right Time

p. 24, "the Devil in human shape" Arthur Koestler, *The Sleepwalkers* (New York: The Macmillan Company, 1959), 127.

Chapter Four: The Return to Italy

p. 53, "the worms with many feet . . ." David C. Knight, *Copernicus, Titan of Modern Astronomy* (New York: Franklin Watts, Inc., 1965), 45.

p. 54, "Armenian sponge, cinnamon, cedar wood . . ." Leopold Prowe, *Nicolaus Copernicus* (Berlin: 1883-4, Vol I, 2), 313.

Chapter Five: Church and State

p. 68, "explores the rapid course of the moon . . ." Knight, *Titan*, 57.

Chapter Six: The Duties of the Canon

p. 79, "The honored highly learned . . ." Knight, *Titan*, 71.

Chapter Seven: The Astronomer Emerges

p. 87, "everything would move uniformly . . ." Edward Rosen, trans., *Three Copernican Treatises* (New York: Octagon Books, 1971), 57.

p. 87, "reserving these for my larger . . ." Ibid., 59.

p. 87, "Then Mercury runs on seven circles . . ." Ibid., 90.

Chapter Eight: *On Revolutions*

p. 92, "bequeathed to us . . ." Rosen, *Three Copernican Treatises*, 99.

Chapter Nine: An Elder Canon

p. 108, "I can well appreciate, Holy Father . . ." A.M. Duncan, trans., *Copernicus: On the Revolutions of the Heavenly Spheres* (New York: Barnes and Noble Books, 1976), 23.

p. 109, "a learned monk who has his own opinions . . ." Koestler, *Sleepwalkers*, 143.

p. 109, "cursed for all eternity and smitten with the sword . . ." Ibid.

p. 110, "Oh, if only the Christian spirit informed the Lutheran attitude . . ." Prowe, *Nicolaus Copernicus*, 143.

p. 114, "Frombork wenches" Koestler, *Sleepwalkers*, 182.

p. 114, "trumped up charges" Prowe, *Nicolaus Copernicus,* 366.

Chapter Ten: Publication

p. 116, "There was no place in science . . ." Rosen, *Three Treatises,* 193.

p. 123, "These hypostheses need not . . ." Edward Rosen, *Copernicus and the Scientific Revolution.* (Malabar, Fla.: Robert E. Krieger Publishing Co., 1984), 195.

p. 124, "For my part I have always felt . . ." Koestler, *Sleepwalkers*, 167.

p. 125, "For many days he had been deprived . . ." Prowe, *Nicolaus Copernicus,* 554.

BIBLIOGRAPHY

Duncan, A.M., trans., *Copernicus: On the Revolutions of the Heavenly Spheres*. New York: Barnes and Noble Books, 1976.

Ferguson, Kitty. *Tycho & Kepler: The Unlikely Partnership that Forever Changed Our Understanding of the Heavens*. New York: Walker and Co., 2002.

Gingerich, Owen. *The Eye of Heaven: Ptolemy, Copernicus, Kepler*. New York: The American Institute of Physics, 1993.

———. *The Great Copernicus Chase*. Cambridge, England: Cambridge University Press, 1992.

Hall, A. Rupert. *From Galileo to Newton*. New York: Dover Publications, Inc., 1981.

Hirshfeld, Alan W. *Parallax: The Race to Measure the Cosmos*. New York: Henry Holt and Co., 2001.

Hoskin, Michael, ed. *The Cambridge Concise History of Astronomy*. Cambridge, England: Cambridge University Press, 1999

Knight, David C., *Copernicus, Titan of Modern Astronomy*. New York: Franklin Watts, Inc., 1965.

Koestler, Arthur, *The Sleepwalkers: A History of Man's Changing Vision of the Universe*. New York: Macmillan, 1959.Koyré,

Alexandre. *The Astronomical Revolution: Copernicus-Kepler-Borelli*. Trans. R.E.W. Maddison. New York: Dover Publications, Inc., 1973.

Kuhn, Thomas S. *The Copernican Revolution: Planetary Astronomy in the Development of Western Thought*. Cambridge, Mass.: Harvard University Press, 1957.

———. *The Structure of Scientific Revolutions*. Chicago: The University of Chicago Press, 1996.

Prowe, Leopold, *Nicolaus Copernicus*. Berlin: 1883-4, Vol I, 2.

Rosen, Edward. *Copernicus and the Scientific Revolution*. Malabar, Fla.: Robert E. Krieger Publishing Co., 1984.

———. *Three Copernican Treatises*. New York: Octagon Books, 1971.

WEBSITES

The Copernican Model
http://csep10.phys.utk.edu/astr161/lect/retrograde/
copernican.html

School of Mathematical and Computational Sciences
University of St Andrews
http://www-groups.dcs.st-and.ac.uk/~history/Mathematicians/
Copernicus.html

Science World: Wolfram Research
http://scienceworld.wolfram.com/biography/Copernicus.html

INDEX